RECENT TRENDS IN BIOTECHNOLOGY

HIGH VALUE PROCESSING TECHNOLOGIES

RECENT TRENDS IN BIOTECHNOLOGY

Additional books in this series can be found on Nova's website
under the Series tab.

Additional e-books in this series can be found on Nova's website
under the e-book tab.

RECENT TRENDS IN BIOTECHNOLOGY

HIGH VALUE PROCESSING TECHNOLOGIES

AYDIN BERENJIAN
HODA JAFARIZADEH-MALMIRI
AND
YUANDA SONG
EDITORS

nova publishers
New York

Library of Congress Cataloging-in-Publication Data

Names: Berenjian, Aydin, editor. | Jafarizadeh-Malmiri, Hoda , editor. | Song, Yuanda, editor.
Title: High value processing technologies / editors, Aydin Berenjian, Hoda Jafarizadeh-Malmiri and Yuanda Song (The University of Waikato, New Zealand, and others).
Description: Hauppauge, New York : Nova Science Publisher's, Inc., [2016] | Series: Recent trends in biotechnology | Includes index.
Identifiers: LCCN 2016000060 (print) | LCCN 2016001329 (ebook) | ISBN 9781634844536 (hardcover) | ISBN 9781634846295 (ebook) | ISBN 9781634846295 ()
Subjects: LCSH: Microbial biotechnology. | Biomolecules.
Classification: LCC TP248.27.M53 H54 2016 (print) | LCC TP248.27.M53 (ebook) | DDC 572--dc23
LC record available at http://lccn.loc.gov/2016000060

Published by Nova Science Publishers, Inc. † New York

CONTENTS

PREFACE

High Value Processing is concerned with all the processes and activities of converting raw materials and commodity materials into valuable products required by manufacturers or the end consumer. Efficient processing of commodity goods is of great significance for continuing economic welfare. There is a need to develop products that have high value in world markets, and therefore must understand fully the properties of materials as diverse as food, wood, metals, plastics pharmaceutical and fuel. Advanced technologies and knowledge are required to design, manufacture and process these materials into high-value products. It is also important to understand the properties of these high-value products and how they will interact with their environment, whether it be within the body or in the atmosphere. This book is the most comprehensive and ambitious undertaking we are aware of on the subject of high value processing in that it covers most of the relevant novel technologies. All chapters are written by key scientists with diverse backgrounds in either industry or academia, and all provide an update on emerging technologies as well as visions for the future. Presently, new processing technologies are capturing the attention of many key scientists in academia and government. Novel processing technologies have become more sophisticated and diverse in response to the growing demand for quality products. Strategies to meet such demands include modification to existing processing techniques and the adoption of novel processing technologies. Innovation is a key factor that can be improved significantly through fundamental research covering a broad spectrum of disciplines prior to the commercialization of new products and technologies. Therefore, in order to meet the demand for better quality products, every effort should be made to understand the basic principles behind processing, as well as to recognize new opportunities and to consider combined strategies. This book will serve industrial and other activities where material is undergoing a change, whether that change is chemical, biochemical or physical. Improving the understanding of how to prepare feed materials, how to make reactions occur, separating and purifying products, controlling wastes, minimising energy usage, and ultimately adding value to the raw materials used to produce something useful to people are essential aspects to this composition. This book reads more as a single authored book with fully integrated chapters than as one compiled by editors, having benefited directly from the discussions by true experts in their fields. The ten chapters of this book are: "Aqueous Two Phase Extraction of Proteins," "Production of Nanodispersions by Solvent-Displacement Technique," "Continuous Manufacturing of Pharmaceutical and Biopharmaceutical Agents," "Producing More with Less," "Supercritical Fluid Extraction of Value Added Compounds from Food Commodities," "Structure,

Synthesis and Environmental Applications of Nano-Sized Hydroxyapatite," "The Effectiveness of Microbial Crack Treatment in Self Healing Concrete," "Preparation and Application of Multi-Enzyme Complexes and Pectic Oligosaccharides from Orange Peel Wastes," "Isolation Process of Antimicrobial Compounds from Plants," "Production of Polyunsaturated Fatty Acids by Oleaginous Microorganisms and Plants," and "Fungal Cellulase Based on Mutant Aspergillus Spices." We strongly believe this book will be embraced by scientists in academia and government as a valuable reference used for consultation on matters of novel processing technologies.

Aydin Berenjian
Hoda Jafarizadeh-Malmiri
Yuanda Song

In: High Value Processing Technologies
Editors: A. Berenjian, H. Jafarizadeh-Malmiri et al.
ISBN: 978-1-63484-453-6

Chapter 1

AQUEOUS TWO PHASE EXTRACTION OF PROTEINS

Tim Zeiner*
TU Dortmund University,
Department of Biochemical and Chemical Engineering,
Laboratory of Fluid Separations, Dortmund, Germany

ABSTRACT

Downstream processes which are based on chromatography are limited due to capacity and scale-up issues. Classical extraction processes which are known from chemical engineering could overcome this capacity problem. Moreover, it is already known that aqueous two-phase systems (ATPS) can be applied for the extraction of biomolecules. ATPS can be formed by mixing two hydrophilic components above a critical concentration. The major advantage of ATPS compared to aqueous-organic systems lies in the aqueous character of both phases which leads to an increasing biocompatibility. In this contribution the application of ATPS for the purification of proteins is shortly reviewed. Moreover, two applications of multistage extraction of ATPS for the separation of enzymes and for the purification of monoclonal antibodies are presented. For both tasks a polyethylene glycol - phosphate salt ATPS was used. As model system for enzymes, the separation of laccase of *Pleurotus Sapidus* from laccase of *Trametes Versicolor* is shown. In addition to the experimental investigation of the process, it was also modelled using an equilibrium stage model. To purify monoclonal antibodies an extraction process is exhibited which compromises an extraction part at high sodium chloride (NaCl) concentration, a back extraction step at low NaCl concentration and a washing part with nearly no NaCl. By this extraction configuration, a high purity of monoclonal antibodies was achieved.

INTRODUCTION

In 1896, Beijerinck discovered the first aqueous two phase system (ATPS) after mixing an aqueous gelatine solution (10 wt.-%) with an aqueous starch solution (10 wt.-%) [1]. The

* E-mail: tim.zeiner@bci.tu-dortmund.de.

starch-rich phase formed the bottom phase in a flask, whereas the gelatine-rich phase was the top phase. This is the first mentioning of an ATPS formed by two polymers. Later, Per-Åke Albertsson accidently discovered a polymer-salt ATPS by using polyethylene glycol (PEG) as a detergent [2]. Albertsson was the first scientist to apply an ATPS as an extraction system for the separation of cell particles and organelles and proteins [3, 4]. Moreover, the formation of various polymer-polymer and polymer-salt ATPSs was described. Furthermore, Albertsson used ATPSs to purify polyelectrolytes, proteins, nucleic acids detergents and larger compounds as cell organelles and membrane vesicles, which are not soluble particles [5]. Afterwards, many research groups investigated the formation and application of ATPSs to purify and separate biomolecules. For example the groups of Kula [6], Zaslavsky [7] and Walter et al. [8] applied ATPSs for enzyme purification. Moreover, Andrews and Asenjo [9], Rosa et al. [10] and Rito-Palomares [11] applied ATPS to purify biopharmaceuticals.

Aqueous Two Phase Systems

ATPSs are formed by the solution of two hydrophilic components if a certain concentration of these components is reached; both phases have an aqueous character [2]. The hydrophilic components are then called phase-forming components (PFCs). A typical phase diagram of two PFCs is shown in Figure 1. The binodal curve connects the top phase (TP), bottom phase (BP) and the critical point (CP) and separates the two phase region from the homogeneous region. The TP, BP and mixing point (MP) are connected by the tie line and illustrates the state of thermodynamic equilibrium because the tie lines connects corresponding phases. Different concentrations of phase-forming components in corresponding phases lead to differing physical properties in each phase causing a partitioning of molecules between corresponding phases. Thus, the purification of a target component can be achieved. ATPSs offer a gentle environment for the purification of biomolecules due to their aqueous character [2].

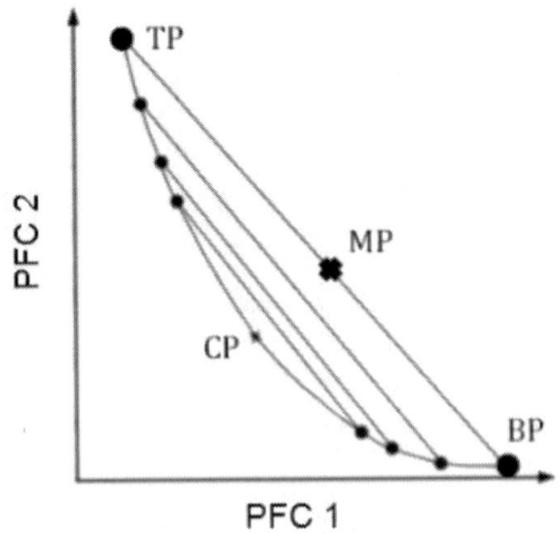

Figure 1. Schematic of the phase equilibrium diagram for an ATPS.

The formation of two phases is thermodynamically preferred because Gibb's free energy is minimized [12]. This concept is valid for all ATPS, which are formed by any phase forming component. There are different possibilities for forming ATPSs. Classical ATPSs consist of either two polymers (e.g., polyethylene glycol (PEG) and dextrane) or a polymer and a salt (e.g., PEG and a phosphate salt) dissolved in water [2]. However, there are many ATPSs known, Wohlfarth [13] mentioned publications including more than 1200 different ATPSs containing polymer that were published between 1957 and 2011. These only include polymer-based ATPS e.g., polymer-polymer, polymer-polyelectrolyte, polyelectrolyte-polyelectrolyte, polymer-salt, polyelectrolyte-salt, polymer-alcohol, polyelectrolyte-alcohol, polymer-carbohydrate and polyelectrolyte-carbohydrate ATPSs. In addition to polymer based ATPS there are alcohol-salt systems [14] and ATPSs based on ionic liquids (IL) [15], including IL-salt, IL-amino acid, IL-carbohydrate and IL-polymer ATPS.

APPLICATION OF AQUEOUS TWO PHASE SYSTEMS

Extensive research concerning the application of ATPSs for the recovery and partial purification of therapeutic and other proteins of industrial interest has been conducted. One intensively discussed example for therapeutic proteins is the purification of monoclonal antibodies (mAbs) by ATPS. Andrews et al. [9] used an ATPS for the purification of mAbs; activated PEG molecules with Protein A or anti-BSA ligands were applied to enhance the affinity of mAbs influencing the partitioning. Because this procedure is very costly, Andrews et al. [16] presented a different approach. They purified murine Immunoglobulin G (IgG) from hybridoma cells with an ATPS based on PEG and phosphate salt. During this process, protein precipitation was minimised by an adjustment of extraction conditions. To further increase the IgG purity and to regenerate the PEG-rich top phase for solvent recycling, a subsequent back extraction step was performed.

Recently, many publications on the ATPE of mAbs have been released by the Aires-Barros group. They have investigated single-stage extractions of pure mAbs and single-stage extraction of mAbs from cell supernatant.

Rosa et al. [17] investigated the purification of an IgG from synthetic protein mixtures using a PEG - phosphate salt (phosphate) ATPS. The reported yield of native IgG in the TP was 97% with a purity of 99%. To further increase the purity and to regenerate the PEG-rich top phase for recycling, an additional back extraction step was performed. This back extraction step was carried out using additional phosphate buffer at a pH of 6, which led to a total recovery of 76% with an IgG purity of 100%. Protein precipitation was also observed at the interface.

This work was extended by Azevedo et al. [18] by applying a PEG - phosphate ATPS to purify IgG from synthetic media, Chinese hamster ovary (CHO) cell supernatant and hybridoma supernatant. They reported a yield of 88% mAbs from CHO cell supernatant with a purification factor of 4.3.

Because phosphate leads to eutrophication of the environment, Azevedo et al. [19] applied a PEG - citrate ATPS to purify mAbs from hybridoma cell supernatant. Until then, citrate had not been applied to purify mAbs. Azevedo et al. [19] analysed the mAb purification for different molecular weights of PEG in combination with citrate and additional sodium chloride (NaCl). An additional back extraction step with an overall weight fraction of 5 weight percent (wt.-%) NaCl resulted in a final mAb recovery of 99% and a purity of 76%. However the primary drawback of this ATPS is the appearance of precipitation. To decrease the protein precipitation, high volume ratios and low salt amounts were suggested.

Platis et al. [20] presented an application of a PEG - phosphate ATPS to purify plant-derived mAbs. They extracted human anti-human immunodeficiency virus mAb 2F5 from tobacco extract. They reported a mAb purity of 95% in the bottom phase. Regarding the total protein content, a purification factor of 3 – 4 was achieved. Moreover, ATPE removed plant-derived compounds as phenolics and toxic alkaloids. Furthermore, they showed that mAbs were fully active after the extraction.

In addition to polymer-salt ATPSs, polymer-polymer ATPS were also applied for the purification of mAbs. Rosa et al. [21] studied the effect of affinity-enhanced PEG on the partitioning of IgG from CHO (Chinese Hamster Ovary) cell supernatant. They examined the effect of functionalised PEG based on the ligand type and ligand amount. The best achieved yield was 93% with a purification factor of 1.9. Compared to a non-functionalised ATPS, this is a 60% increase in selectivity.

Ferreira et al. [22] purified mAbs from CHO cell supernatant using a thermoseparating ATPS. This ATPS was composed of PEG, dextran and water; IgG1 was primarily partitioned into the dextran-rich phase. Purification of IgG in the top phase was achieved after the addition of triethylene glycol-diglutaric acid with mAb-purity of 88%.

In addition to the therapeutic proteins industrial-relevant proteins such as enzymes have also been purified. Because of the large number of different enzymes purified by ATPE is huge, only selected examples are presented.

α-Amylase is widely used, particularly in the food industry. Bacterial α-amylase is added to barley in breweries to improve the liquefaction of the starch. Schmidt et al. [23] examined the partitioning of α-amylase in different ATPSs (PEG-dextrane, PEG-phosphate and PEG-sulphate systems). They found that the addition of NaCl had the largest impact on the partition coefficient of α-amylase leading to a 1,000-fold increase in the partitioning coefficient for PEG-sulphate, 150-fold increase forPEG-phosphate and 8-fold increase in PEG-dextran systems. However, the best partitioning performance was obtained in the PEG-phosphate system.

Alcohol dehydrogenases are enzymes that occur in many organisms and facilitate the interconversion between alcohols and aldehydes or ketones. In animals, they serve to break down alcohols that would otherwise be toxic; they also participate in generation of useful aldehyde, ketone, and alcohol groups. Moreover, alcohol dehydrogenases can be used to catalyse the breakdown of fuel for an ethanol fuel cell. Szlag and Guiliano [24] used a PEG–maltodextrin system for the purification of alcohol dehydrogenase. After functionalisation of 4% PEG with Cibacron blue, the partition coefficient at low pH-values increased by a factor of eight compared to the unfunctionalised system.

Xylan is an abundant biopolymer that us found in plant tissues and is a major component of the cell wall hydrolysed by xylanase. The large variety of potential applications of these enzymes is the main reason for investigating fungal and bacterial production. Currently, the most important application of xylanases is in prebleaching of kraft pulp. Bim and Franco [25] used an ATPS composed of PEG, phosphate and NaCl. They achieved a purification factor of 33% and a 98% enzyme yield with this ATPS.

β-Galactosidase is an enzyme that is present in microorganisms, plants and animals. Its application in the food industry includes the products, such as sweet syrups, ice creams, dairy desserts, and confectionery of low lactose content and/or increased sweetness. β-Glucosidase is a type of hydrolyase that occurs in various microorganisms, plants and animals tissues. The major applications of β-glucosidases include the production of glucose and other soluble sugars from cellulosic wastes in the process of biomass conversion. Hemavathi and Raghavarao [26] applied an ATPS to separate and purify β-galactosidase and β-glucosidase from barley (*Hordeum vulgare*). Suitable for this purification is an ATPS consisting of PEG and ammonium sulphate. Single-stage extraction resulted in an activity recovery of 98.3% with a purification increase of 2.1 fold for β-galactosidase and an activity recovery of 92.6% with purification of 3.3 fold of β-glucosidase; β-galactosidase and β-glucosidase were selectively partitioned to PEG and sulphate phase, respectively.

These prior investigations were all conducted using a single-stage extraction; however, multistage extraction processes can result in higher purities and yields.

Multistage Extraction Processes

Because most downstream process unit operations, especially chromatography, are conventionally conducted in a discontinuous manner, there is a need for continuous purification unit operations. The major advantage of continuous aqueous two phase extraction (ATPE) is the time and cost reduction. Furthermore, multistage extraction in a countercurrent mode requires lower throughput and raw material consumption [10]. Espitia-Saloma et al. [27] provided an overview of processes and apparatuses applied for ATPE. However, only a few publications focused on that topic have been published. They reported that commonly used extraction devices include column contactors, mixer-settler units (MSUs) and other contactors, such as raining bucket contactor. Because ATPE uses liquid-liquid demixing like extraction process in chemical industry, most standard extraction equipment can be applied for this extraction process [28]. However, density differences between the two phases and the interfacial tensions are lower in ATPSs compared to organic-aqueous systems. While ATPE requires a low energy input to achieve intense phase dispersion [Kul99]. These properties also limit the operating window of a continuous, multistage ATPE because there is a higher probability of flooding, back mixing and emulsification than in classic organic-aqueous systems. These effects can lead to a bad extraction performance [27]. Another limiting factor is the protein solubility in ATPSs [9, 17, 18]. The precipitation of proteins strongly depends on the weight fraction and the type of PFC [29]. To avoid precipitation, operating conditions close to the critical point must be used to ensure high protein solubility by accepting low differences in density and interfacial tension which implies that the handling of a multistage process may be difficulties. Regarding the available extraction apparatuses, a MSU can compensate for these drawbacks because longer settling and mixing times are provided. Furthermore, it is operated regardless of the phase properties and the scale-up is straightforward. Additional advantages include high reliability, high flexibility and capacity. However, the drawbacks are higher investment costs as well as operational costs compared to other continuous devices.

The first reported application of a MSU in an ATPE process was by Veide et al. [30]. They combined a static mixer with a subsequent centrifuge for phase separation to extract β-galactosidase from E. coli using a PEG-phosphate ATPS.

Rosa et al. [10] were the first to apply a continuous multistage ATPE process in a MSU to purify IgG from cell supernatant in a PEG-phosphate ATPS. They developed a process comprising an extraction step to extract mAbs into the PEG-rich top phase by adding NaCl to the ATPS. This step was followed by a back extraction of mAbs into the phosphate-rich bottom phase by decreasing the sodium chloride (NaCl) concentration. Afterwards, the bottom phase was washed with the fresh phase PEG-rich phase, while the mAbs remained in the bottom phase. This concept was described by Hustedt et al. and by Kula et al. [31, 28]. Furthermore, they suggested recycling the top phase from the back extraction to reduce raw material consumption.

In the work of Rosa et al. [10], continuous ATPE experiments were conducted with up to ten extraction stages, one back extraction and three washing stages. They achieved a final purity of up to 97% with an IgG yield of up to 80%. Based on their work, they have demonstrated that multistage ATPE can be an economically favorable alternative to purify IgG from cell supernatant. Furthermore, they reported the economical effectiveness of a downstream process based on ATPE [32] and compared it to the current downstream process based on Protein A chromatography.

Espitia-Saloma et al. [27] noted that it is essential to surpass the unattractive economical image of an ATPE process. There is a lack of designing rules for effective process development, which limits the application at industrial scale. Therefore, guidelines for design/scale up, performance models and the investigation of phase recycling are required.

Rito-Palomares and Lyddiatt [33] examined the recycling of PEG-rich top phase from the back extraction step of a two-stage ATPS for the extraction of bulk protein, pyruvate kinase and fumarase. Via partially recycling the top phase, the raw material consumption was reduced with only minor effects on the extraction performance. Recycling of the top phase from the back extraction into the extraction step was also found to reduce the raw material consumption in the multistage extraction. In the following chapters two examples for the development of a multistage extraction process are shown.

Multistage Extraction of Monoclonal Antibodies

The first step in the development of multistage processes is the investigation of the extraction system. Here, the protein solubility in an ATPS is a significant limitation to obtain a good extraction performance of mAbs from a cell supernatant [29]. Supporting the ATPS choice by a solubility screening of the target component is a helpful tool to determine promising extraction conditions. Using a PEG-phosphate system, mAbs can either be purified in the top- or in the bottom phase, which depends on the weight fraction of NaCl present in the ATPS. High weight fractions above 10 wt.-% NaCl and low weights fractions below 4 wt.-% NaCl delivered the most promising extraction conditions. A decrease of product phase volume also had a positive effect on the extraction performance. The partition behaviour of different mAbs types in different PEG – phosphate ATPS is thus comparable, although the ATPS conditions differ with regard to PEG molecular weight and phase composition. This simplifies system characterization and lead to a generalized approach to determine sufficient

ATPS for the extraction of mAbs from cell supernatant. To achieve a high final purity of mAbs, it was necessary to apply several separation stages. Therefore, the ATPE of mAbs from cell supernatant had to be carried out at a multistage scale.

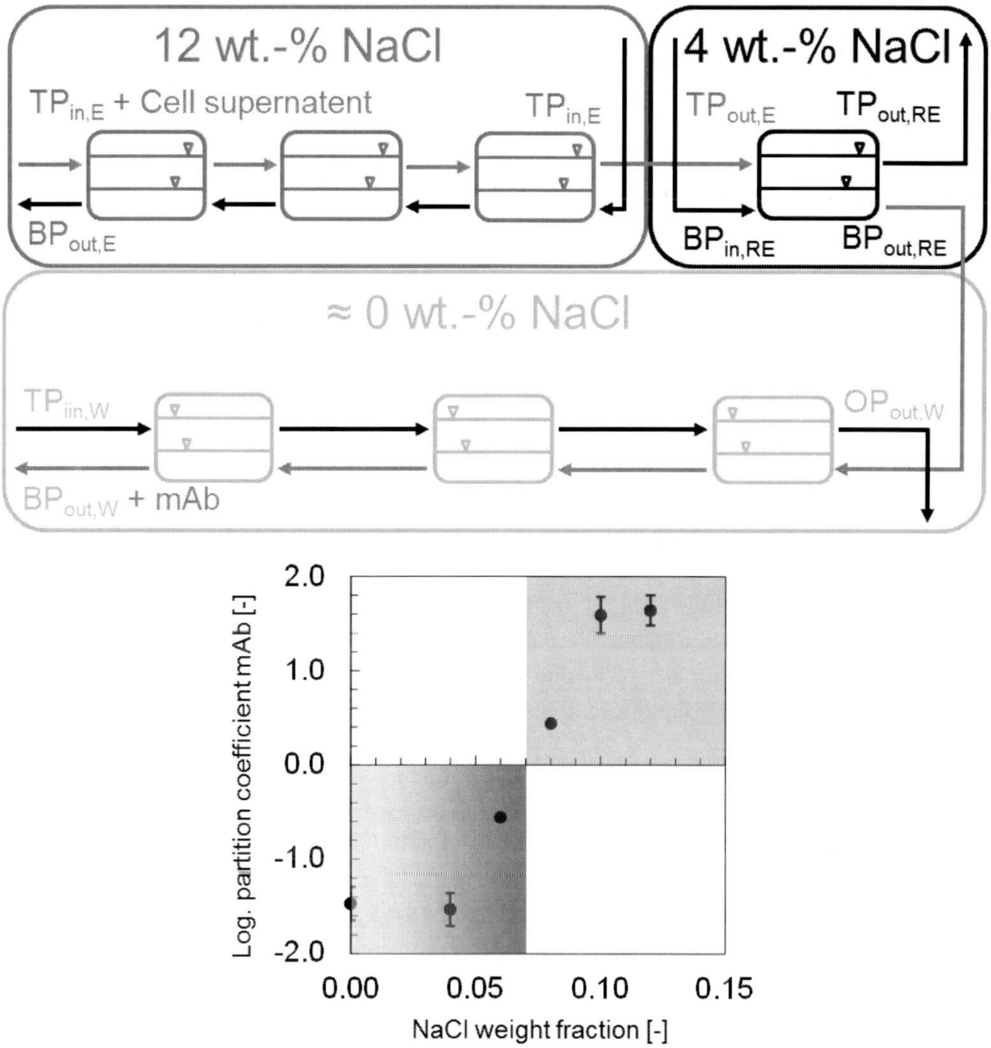

Figure 2. Top: Extraction process for mAbs with a extraction step (E), back extraction step (BE) and a washing step (W). Bottom: Partitioning of mAbs based on sodium chloride weight fraction.

Continuous multistage experiments without NaCl, performed as a washing step, showed that an increase of the number of stages and a decrease of phase ratio have a positive effect on the extraction performance with yields up to nearly 90% [34]. For continuous multistage experiments with NaCl, performed as a stripping step, extremer partition conditions have to be adjusted by adding 12 wt.-% of NaCl to the ATPS. From multistage operation experiments it was observed, that depending on the amount of NaCl added to the ATPS, different molecular weight impurities could be removed from IgG1. Therefore, the combination of a salt-rich extraction and a salt-poor back extraction is a promising option to achieve an improved extraction performance. Nonetheless, mAb-yield losses hint at the limited loading capacity of ATPS to dissolve mAbs and require high amounts of phase-forming components. Phase recycling is a promising method to decrease the raw material consumption.

For processes applying in total seven stages (Figure 2) and a low cell supernatant loading, high IgG1 yields of 81.3 – 88.8% and purities in the range of 93.4 – 95.8% were achieved [35]. Recycling of 40 wt.-% top phase from the back extraction step into the extraction step was proven as acceptable amount for the recycle stream in equilibrium experiments over ten recycles. During continuous multistage ATPE phase recycling did not have an impact on the extraction performance and led to a significant reduction of raw material consumption. The limited loading capacity of the used ATPS was also found to be the major drawback during the continuous, multistage ATPE. The precipitation caused a significant IgG1 loss during the process. With regard to the scale up of the extraction process a tremendous amount of phase forming components would be required [35].

Multistage Extraction of Laccase

The system of Prinz et al. [36, 37] which comprised laccase from *Pleurotus sapidus* and *Trametes versicolor* was selected to investigate the separation of two similar enzymes by aqueous two-phase extraction. Therefore, a culture supernatant of a *Pleurotus sapidus* fermentation was spiked with laccase from *Trametes versicolor*. Thus the total activity of the enzyme was doubled. Single-stage experiments revealed a strong influence of sodium chloride on the activity partitioning of laccase from *Pleurotus sapidus* in PEG3000-phosphate ATPS buffered at pH 7 and 25°C. The addition of the salt increased the activity partitioning of both laccases into the bottom phase. Because, laccase of *Pleurotus sapidus* partitioned into the top phase in ATPS without sodium chloride, the salt was used as a "switch" such as in the case of mAbs. In the course of these experiments it was shown, how clearance factors (is applied for the assessment of the separation of one laccase from the other) for mixtures can be calculated from two superimposed experiments with only one type of laccase.

The addition of sodium chloride enabled a purification of laccase from both hosts in the bottom phase with a purification factor of 2.10 and an activity recovery in the phase of 83 wt-%. In contrast, systems without sodium chloride enabled a separation of the laccase from *Trametes versicolor* in the bottom phase with a clearance factor of 6.35 and an enzyme specific activity recovery of 122% (a recovery of more than 100% can occur, when the activity is increased by the PFC). Simultaneously, laccase from *Pleurotus sapidus* is separated in the top phase with a clearance factor of 3.34 and an enzyme specific activity recovery of 10%. In this case, the addition of sodium chloride increased the clearance factor of laccase from *Pleurotus sapidus* to 9.71, whereas the activity recovery decreases to 49%.

After a single-stage investigation of the ATPE, a multi-stage extraction model based on an equilibrium approach was proposed by Prinz et al. [36]. The model calculated mass and volume flows, mass fractions of phase forming components as well as enzyme activities on every stage in every phase of the extraction process. Moreover, the model considered the activity in- and decreasing effect of the ATPS on the enzyme activity. Input variables for solving the model equations were the mass fractions of phase forming components as well as mass flows and enzyme activities in the entering streams. The parameters for the model equations were fitted from data available to single-stage experiments.

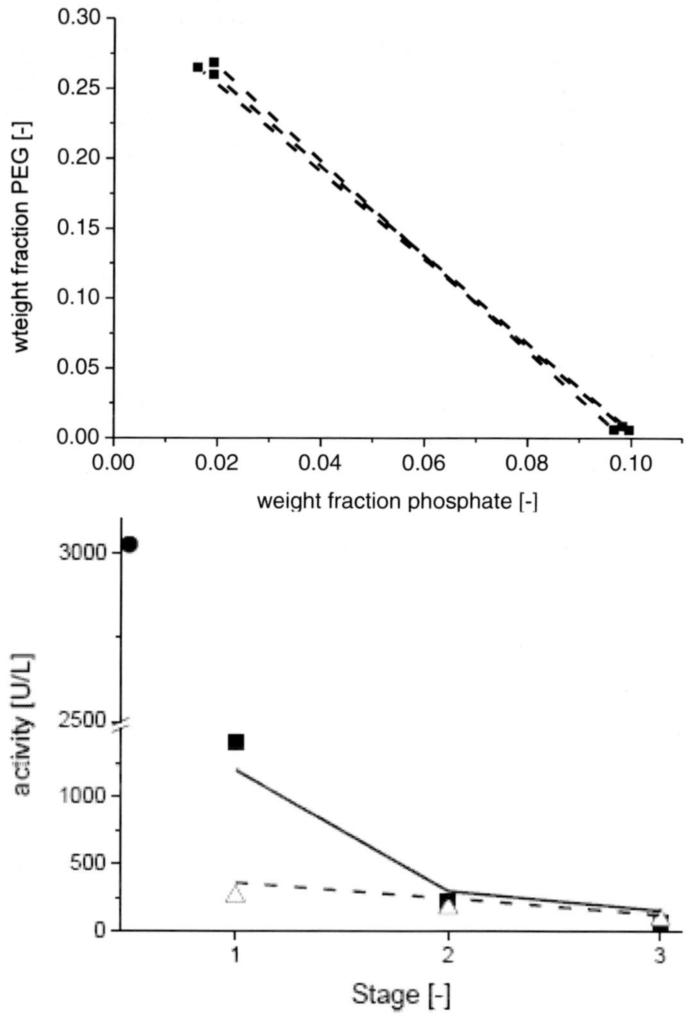

Figure 3. Experimental and modeling results for purification of *Pleurotus sapidus* and *Trametes versicolor* laccases fed in the top phase in a PEG3000-phosphate ATPS at pH 7 and 25°C with 2.5 wt-% sodium chloride: Top: Mass fraction of PFCs on different stages; Bottom: laccase activity in the feed (full circle) and top (mod: line; exp: filled squares) and bottom phase (mod: broken line; exp: open triangles) [36].

Based on the results from single-stage extractions continuous multistage extraction experiments were conducted in a three-stage mixer-settler miniplant. By this change from batch to continuous, single- to three-stage and lab- to miniplant scale, purification factors and activity recoveries were significantly increased. The best process configuration for the separation of an equal mixture of laccase from *Pleurotus sapidus* and from *Trametes versicolor* comprised a feed in the top phase and an aqueous two-phase system with sodium chloride. The purification factor in the bottom phase of this process was 2.84 with an activity recovery of 104%. By comparing the results of the experiments with corresponding modeling results, the proposed model was evaluated (Figure 3). Relative deviations between experimental and modeled activity recovery and partitioning were below 15% with very few exceptions. In addition, the relative deviations between modeled and measured mass fractions of phase forming components were generally below 5%. Therefore, the proposed equilibrium based model was considered as valid. In systems without sodium chloride, the clearance factor of laccase from *Pleurotus sapidus* in the top phase is 64.3, whereas 8% of the activity of laccase from *Pleurotus sapidus* was recovered. A clearance factor of 6.0 was achieved for laccase from *Trametes versicolor* in the bottom phase with an enzyme specific activity recovery of 122%.

CONCLUSION

This work summarised applications of aqueous two phase extraction for the purification of different biomolecules. As there are a huge number of different possible aqueous two phase systems (ATPS) there is still a high experimental effort to find an appropriate ATPS for a separation task. However, high purity and yield can be reached by using a not optimal ATPS. Moreover, the effort for process development can be minimised by the use of a process model allowing designing the process based on single stage experiments.

REFERENCES

[1] Beijerinck, M. W., (1910). Über Emulsionsbildung bei der Vermischung wässriger Lösungen gewisser gelatinierender Kolloide. *Colliod and Polymer Science,* 7, 16–20.

[2] Albertsson, P., (1990). Separation of cell particles and molecules: A citation-classic commentary on partition of cell particles and macromolecules by Albertsson, P. A.. Current Contents/Life Sciences, 1990, 22.

[3] Albertsson, P. (1958). Particle fractionation in liquid two-phase systems: The composition of some phase systems and the behaviour of some model particles in them application to the isolation of cell walls from microorganisms. *Biochimica et Biophysica Acta,* 27, 378–95.

[4] Albertsson. P., (1958). Partition of proteins in liquid polymer-polymer two-phase systems. *Nature*, 182, 709–11.

[5] Albertsson, P., (1986) Partition of cell particles and macromolecules. 3rd ed. New York: John Wiley and Sons.

[6] Kroner, K. H., Hustedt, H., Granda, S., Kula, M. R., (1978). Technical aspects of separation using aqueous two-phase systems in enzyme isolation processes. *Biotechnology and Bioengineering*, 20, 1967–88.

[7] Zaslavsky, B. Y., (1995). Aqueous two-phase partitioning: Physical chemistry and bioanalytical applications. New York NY u. a: Dekker.

[8] Walter, H., Fisher, D., Brooks, D. E., (1985). Partitioning in aqueous two-phase systems: Theory, methods, uses, and applications to biotechnology. Orlando: Academic Press.

[9] Andrews, B. A, Asenjo J. A., (1996). Protein partitioning equilibrium between the aqueous poly(ethylene glycol) and salt phases and the solid protein phase in

poly(ethylene glycol)-salt two-phase systems. *Journal of Chromatography B: Biomedical Sciences and Applications*, 685, 15–20.

[10] Rosa, P. A. J., Azevedo, A. M., Sommerfeld, S., Mutter, M., Bäcker, W., Aires-Barros, M. R. (2013). Continuous purification of antibodies from cell culture supernatant with aqueous two-phase systems: From concept to process. *Biotechnology Journal*. 8, 352-62.

[11] Rito-Palomares, M., (2004). Practical application of aqueous two-phase partition to process development for the recovery of biological products. *Journal of Chromatography B*, 807, 3–11.

[12] Flory, P. J., (1978). Principles of polymer chemistry. 10th ed. Ithaca, N.Y: Cornell Univ. Pr.

[13] Wohlfarth, C., (2012). CRC handbook of phase equilibria and thermodynamic data of aqueous polymer solutions. Boca Raton, FL: Taylor and Francis.

[14] Tianwei, T., Qing, H., Qiang, L., (2002). Purification of glycyrrhizin from Glycyrrhiza uralensis Fisch with ethanol/phosphate aqueous two phase system. *Biotechnology Letters*. 24, 1417-1420.

[15] Freire, M. G., Cláudio, A.F.M., Araújo, J. M. M., Coutinho, J.A.P., Marrucho, I. M., Lopes, J.N.C., (2012). Aqueous biphasic systems: a boost brought about by using ionic liquids. *Chemical Society Reviews*, 41, 4966–95.

[16] Andrews, B. A., Asenjo, J. A., (1996) Protein partitioning equilibrium between the aqueous poly(ethylene glycol) and salt phases and the solid protein phase in poly(ethylene glycol)-salt two-phase systems. *Journal of Chromatography B: Biomedical Sciences and Applications*, 685, 15–20.

[17] Rosa, P. A. J., Azevedo, A. M., Aires-Barros, M. R.,(2007). Application of central composite design to the optimisation of aqueous two-phase extraction of human antibodies. *J. Chromatogr. A*., 1141, 50-60.

[18] Azevedo, A. M., Rosa, P. A. J., Ferreira, I. F., Aires-Barros, M. R., (2007). Optimisation of aqueous two-phase extraction of human antibodies. *J. Biotechnol*., 132, 209-217.

[19] Azevedo, A. M., Gomes, A. G., Rosa, P. A. J., Ferreira, I. F., Pisco, A. M. M. O., Aires-Barros, M. R., (2009). Partitioning of human antibodies in polyethylene glycol-sodium citrate aqueous two-phase systems. *Sep. Purif. Technol*., 65, 14-21.

[20] Platis, D., Labrou, N. E., (2006). Development of an aqueous two-phase partitioning system for fractionating therapeutic proteins from tobacco extract. *J. Chromatogr. A*., 1128, 114-24.

[21] Rosa, P. A. J., Azevedo, A. M., Ferreira, I. F., de Vries, J., Korporaal, R., Verhoef, H. J., Visser, T. J., Aires-Barros, M. R., (2007). Affinity partitioning of human antibodies in aqueous two-phase systems. *J. Chromatogr. A*., 1162: 103-13.

[22] Ferreira, I. F., Azevedo, A. M., Rosa, P. A. J., Aires-Barros, M. R., (2008). Purification of human immunoglobulin G by thermoseparating aqueous two-phase systems. *J. Chromatogr. A*., 1195, 94-100.

[23] Schmidt, A. S., Ventom, A. M., Asenjo, J. A. (1994). Partitioning and purification of α-amylase in aqueous two-phase systems. *Enzyme and microbial technology, 16, 131-142.*

[24] Szlag, D. C., Guiliano, K. A., (1988). A low-cost aqueous two phase system for enzyme extraction. *Biotechnology techniques*, 2, 277-82.

[25] Bim, M. A., Franco, T. T., (2000). Extraction in aqueous two-phase systems of alkaline xylanase produced by Bacillus pumilus and its application in kraft pulp bleaching. *Journal of Chromatography B: Biomedical Sciences and Applications*, 743, 349-56.

[26] Hemavathi, A. B., Raghavarao, K. S. M. S., (2011). Differential partitioning of β-galactosidase and β-glucosidase using aqueous two phase extraction. *Process Biochemistry*, 46, 649-655.

[27] Espitia-Saloma, E., Vázquez-Villegas, P., Aguilar, O., Rito-Palomares, M., (2014). Continuous aqueous two-phase systems devices for the recovery of biological products. *Food Bioprod. Process*, 92, 101-12.

[28] Kula, M. R., Selber, K., (1999). Protein purification, aqueous liquid extraction. In: Flickinger MC, Drew SW, eds. The encyclopedia of bioprocess technology: Fermentation, biocatalysis, and bioseparation. Vol. 4. New York: Wiley.

[29] Muendges, J., Stark, I., Mohammad, S., Górak, A., Zeiner, T. (2015). Single stage aqueous two-phase extraction for monoclonal antibody purification from cell supernatant. *Fluid Phase Equilibria*, 385, 227-36.

[30] Veide, A., Lindbäck, T., Enfors, S.-O., (1984). Continuous extraction of β-d-galactosidase from Escherichia coli in an aqueous two-phase system: effects of biomass concentration on partitioning and mass transfer. *Enzyme Microb. Technol.*, 6, 325-30.

[31] Hustedt, H., Kroner, K. H., Menge, U., Kula, M. R., (1985). Protein recovery using two-phase systems. *Trends Biotechnol.*, 3, 139-44.

[32] Rosa, P. A. J., Azevedo, A. M., Sommerfeld, S., Bäcker, W., Aires-Barros, M. R., (2011). Aqueous two-phase extraction as a platform in the biomanufacturing industry: Economical and environmental sustainability. *Biotechnol. Adv.*, 29, 559-567.

[33] Rito-Palomares, M., Lyddiatt, A., (1996). Impact of cell disruption and polymer recycling upon aqueous two-phase processes for protein recovery. J. Chromatogr. *B Biomed. Appl.*, 680, 81-89.

[34] Muendges, J., Zalesko, A., Górak, A., Zeiner, T., (2015) Multistage aqueous two-phase extraction of a monoclonal antibody from cell supernatant. *Biotechnology progress*, 31, 925-936.

[35] Muendges, J., Zierow, J., Zeiner, T., (2015) Experiment and Simulation of an Aqueous Two-Phase Extraction Process for the Purification of a Monoclonal Antibody. *Chemical Engineering and Processing: Process Intensification*, 95, 31-42.

[36] Prinz, A., Hönig, J., Schüttmann, S., Zorn, H., Zeiner, T., (2014). Separation and purification of laccases from two different fungi using aqueous two-phase extraction. *Process Biochemistry*, 49, 335-346.

[37] Prinz, A., Koch, K., Gorak, A., Zeiner, T., (2014). Multi-stage laccase extraction and separation using aqueous two-phase systems: Experiment and model. *Process Biochemistry*, 49, 1020-31.

In: High Value Processing Technologies
Editors: A. Berenjian, H. Jafarizadeh-Malmiri et al.
ISBN: 978-1-63484-453-6

Chapter 2

PRODUCTION OF NANODISPERSIONS BY SOLVENT-DISPLACEMENT TECHNIQUE

Navideh Anarjan[1,]*, Mohammad Ali Ghaz-Jahanian[2], Zahra Sayyar[2], Mojgan Mohamadlou[2], Sharareh Najafi[2], Zahra Pazhohnia[2] and Naghmeh Jaberi[1]

[1]Department of Chemical Engineering, Faculty of Engineering, Tabriz Branch, Islamic Azad University, Tabriz, Iran
[2]Faculty of Chemical Engineering, Sahand University of Technology, Tabriz, Iran

ABSTRACT

Functional lipid compounds have recently been received great attention due to their antioxidant, anti-inflammatory, wound healing, and even anti-cancer effects. However, they usually suffer from poor water solubility, low oral absorption, chemical instability, and poor bioavailability. Incorporation of these compounds into nanosized- delivery systems such as nanodispersions can be a good solution for stated problems. An increased various types of nanodispersions are being established for food, agrochemical and pharmaceutical applications in which these are used to encapsulate, protect, and release hydrophobic active compounds. Solvent-displacement or nano-precipitation technique is the most proficient, easy and cost-effective techniques for preparation of nanodispersion systems. This chapter is focused on the nanoparticle creation mechanisms, classifications and important parameters of this technique controlling the characteristics of produced water dispersible nanoparticles and leading to production of most desirable nanodispersions with the smallest particle size, polydispersity index and highest chemical and physical stabilities.

Keywords: functional lipid bioactive compounds, solvent displacement, nanoprecipitation, nanodispersions, nanoparticles

* Corresponding author at: Department of Chemical Engineering, Faculty of Engineering, Tabriz Branch, Islamic Azad University, Tabriz, Iran, Tel: +98-413-3300799, Fax: +98-413-3444355, E-mail: anarjan@gmail.com, anarjan@iaut.ac.ir.

Introduction

It was reported more than 40% of recently discovered drugs in the pharmaceutical and about 70% of new biological functional compounds have poor water solubility and consequently, low bioavailability [1, 2]. Therefore, most of poorly water-soluble components are problematic in use because of their less dissolution rates in biological fluids, and consequently, their limited absorption into cells. Formulate the drugs as nanoparticles or nanodispersions (with the diameters less than 1 μm) can be an attractive approach to improve the dissolution rates of poorly water-soluble compounds [1-3]. The solvent displacement or nano-precipitation technique, based on the impulsive emulsification of the organic phase containing the lipophilic compound, polymer and suitable organic solvent into the aqueous phase containing emulsifier or water soluble polymer, was first developed by Fessi and co-workers [4] in order to produce the water insoluble compounds in drug delivery uses [5, 6]. Nanodispersions hold numerous advantages in comparison to lipid-based solubilization techniques for improving the dissolution rates of compounds. This economical particle formation procedure is more amenable to both water and oils insoluble compounds [1, 7]. The precipitation process is a combination of two steps, namely, nucleation and growth. Nucleation step plays an important role in the characteristics of the final product. Nucleation process is mostly completed in a short time, then the growth process, controlling by either diffusion or surface reaction, is started [8, 9]. Several types of nucleation can be occurred due to absence or presence of foreign particles, i.e., homogeneous or heterogeneous nucleation, respectively [10]. This chapter will focus on solvent displacement or nano-precipitation process principles, main parameters and applications in preparation of functional lipid nanodispersion and nanoparticles.

Mechanisms of Nucleation and Growth of Particles

In nucleation process, the atoms (or ions) that are free in solution gather together to produce a thermodynamically stable cluster. The cluster must exceed a specific size determined by the competition between the Laplace pressure and the free energy supporting the growth of the new phase [3, 8, 9]. LaMer's suggested the mechanism of this stage schematically using a simple diagram shown in Figure 1. As evident from LaMer's diagram, for mono dispersity, high nucleation rate resulting the formation of nuclei in a short time period, an initial high rate of growth of these nuclei to rapidly decrease the concentration less than nucleation concentration and an final slow rate of growth are required for production of fine mono disperse nanoparticles [11].

As shown in LaMer's diagram, the nucleation occurs between the concentrations C*max and C*min. The growth of the particles by diffusion causes the later concentration decreases. The particles will grow until solubility concentrations.

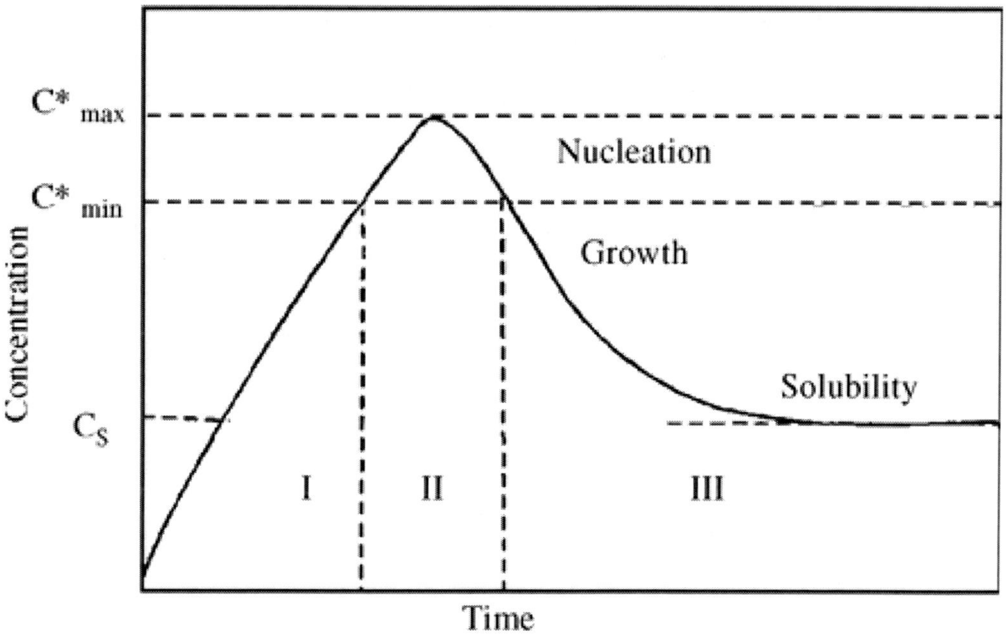

Figure 1. LaMer's diagram (Concentration vs. Time).

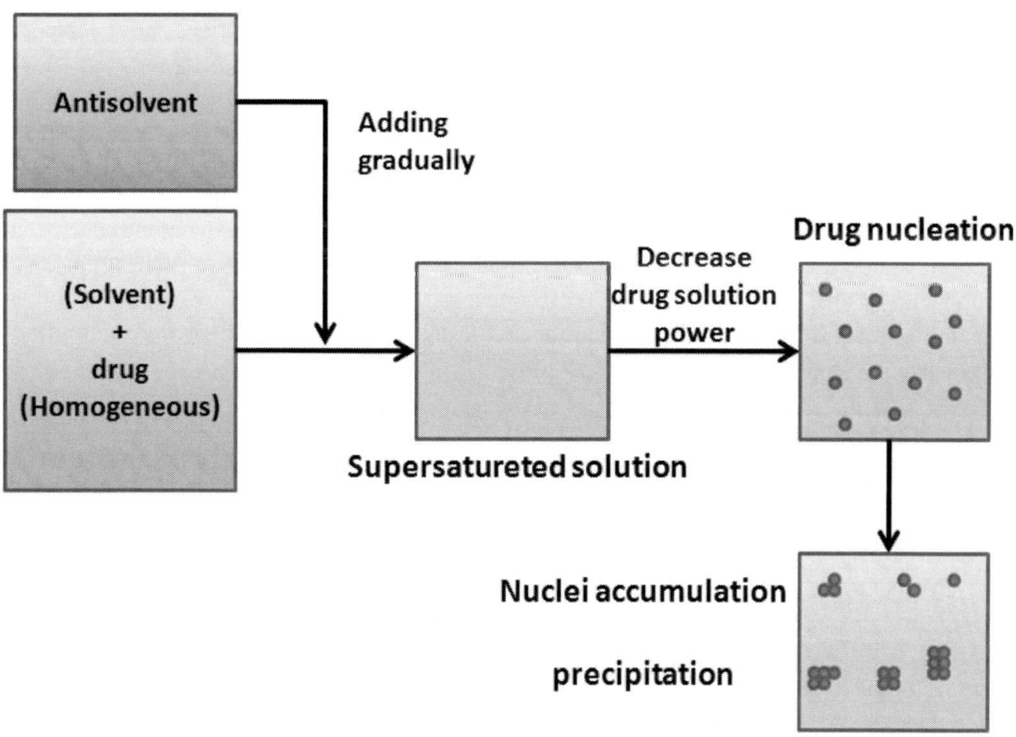

Figure 2. Schematic of particle precipitation process.

No new nuclei are formed as the solute concentration reaches under the critical supersaturation concentration and then nuclei grow by either condensation or coagulation mechanisms. Since the total number and presented surface area of particles are decreased by coagulation, the rate of condensation is reduced. The size of the particles in a solution may also change due to Ostwald ripening. Ostwald ripening, which is controlled either by diffusion or mass transport, is important in systems where the solubility of the solute in the solvent is relatively high [9, 11]. Particle coagulation, which occurs when the attractive interaction dominate the repulsive ones, is another particle growth mechanisms when the solute concentration is lower than the equilibrium saturation concentration. Particles can be stabilized against coagulation using stabilizing agents in the production process [10].

LSW (Lifshitz–Slyozov–Wagner) Model is the most usual theory for nucleation and growth kinetics of the nanoparticles. In this model, the considered parameters are active compound concentration, temperature, diffusion coefficient and the surface energy [8]. Based on the LSW model, the growth behaviour of nanoparticles is modelled involving bulk diffusion and surface reaction process [12]. Spherical shape of particles in a supersaturated medium, the conserved total mass of the solute and insignificance of introducing newly particles processes such as nucleation and aggregation, are the main assumptions of LSW model.

According to LSW, to model the growth step by diffusion mechanism, Fick's first law, can be applied (equation 1), where J, D, and C are the total flux of components passing through a spherical plane with radius x, the diffusion coefficient and the concentration at a distance x, respectively.

$$J = 4\pi x^2 D \frac{dC}{dx} \qquad \text{(Equation 1)}$$

since J is constant and irrespective of x due to the steady state of the solute diffusion, integration of C(x) from (r + δ) to r gives equation 2, for the case of a nanoparticle within solution where δ is the distance from the particle surface to the bulk concentration of components within solution, Cb is the bulk concentration of components within the solution, Ci is the concentration of components at the solid/liquid interface, and Cr is the solubility of the particle,

$$J = \frac{4\pi D r(r+\delta)}{\delta}(C_b - C_i) \qquad \text{(Equation 2)}$$

$$\text{Atoms in a crystal} = \frac{4\pi r^3}{3V_m}$$

$$J = \frac{4\pi r^2}{V_m} \times \frac{dr}{dt}$$

Rate of species entering,

$$\frac{dr}{dt} = \frac{JV_m}{4\pi r^2}$$ (Equation 3)

where *dr/dt* and V_m are growth rate and molar volume respectively. Also

$$J = 4\pi r^2 \times K_d (C_A - C_R)$$ (Equation 4)

$$C_A = \frac{J}{4\pi r^2 K_d} + C_R$$ (Equation 5)

where, K_d and C_R first order decomposition rate constant and solubility of the particle with radius r, respectively. At steady state two fluxes (flux due to diffusion and reaction at the surface) are same. Substituting the value of J, *dr/dt* will be obtained (equation 6):

$$\frac{dr}{dt} = \frac{D(1+\frac{r}{\delta})(C_B - C_R)V_m}{r(1+\frac{D}{rK_d}(1+\frac{r}{\delta}))}$$ (Equation 6)

According to Gibb's-Thompson equation:

$$C_R = C_\infty e^{(\frac{2V_{m\sigma}}{rRT})}$$ (Equation 7)

where, C_∞ is equilibrium concentration at the flat surface (Bulk solubility); σ is interfacial energy; R is universal gas constant and T is temperature. This is a diffusion limited growth mechanism. So for $\delta << r$ and $D << rKd$ equation:

$$\frac{dr}{dt} = K_D \times \frac{(\frac{r}{r_B} - 1)}{r^2}$$

$$K_D = \frac{2V_m^2 D\sigma C_\infty}{RT} \quad (Constant)$$ (Equation 8)

If the total mass of the system is kept conserved, the LSW theory showed that the ratio r/r_B is constant, therefore, equation 8 can be reduced to equation 9 [8, 9]:

$$\frac{dr}{dt} = \frac{K_C}{r^2}$$ (Equation 9)

$$\int_{r_0}^{r} r^2 dr = K_C \int_{0}^{t} dt$$ (Equation 10)

$$\frac{1}{3}(r^3 - r_0^3) = K_C t$$ (Equation 11)

Thermodynamic Aspects

The solute and solvent are completely miscible at the critical temperature or composition area. However, below the upper or above the lower critical solution temperature a miscibility gap arises in metastable area, where phase separation can occur and composition or temperature for a mixture leave the thermodynamically stable range. Metastable is surrounded by the binodal and spinodal curves. In combination with the high solubility of the metastable precipitate, this leads to destabilization issues due to Ostwald ripening, agglomeration and crystallization of the particles. Nucleation is a phase transition were strong changes in composition take place, but which has a small longitudinal range, classically resulting in the formation droplets/particles of well-defined size. Further shifting a sample across the binodal curve leads to increased destabilization of the system and probably the increase of nucleation. For production of homogeneous and high rate of nucleation, the metastable zone should be narrow (Figure 3) [7].

Addition of the solute solution to the anti-solvent or evaporation of solvent decreases the solvent power and causes the supersaturation and nucleation of solute particles. The dispersions obtained by anti-solvent precipitation showed an extreme increased solubility compared to the crystalline solute. Based on the mathematical model developed, further investigations on the details of the phase separation process became possible [7, 13]. Forces between the particles in colloidal systems will be evaluated in next sections.

Mathematical Model

Classical Theory

In classical theory of homogeneous nucleation it is supposed that the critical nuclei are spherical and the structure of the nuclei and their surface can be described by the consistent macroscopic phase [10]. Supersaturation of a solute solution yields nucleation and growth under controlled conditions to influence particle formation. Supersaturation (S) can be expressed as the solute concentration (C) divided by the equilibrium solute solubility (Ceq) under specific conditions [13, 14]:

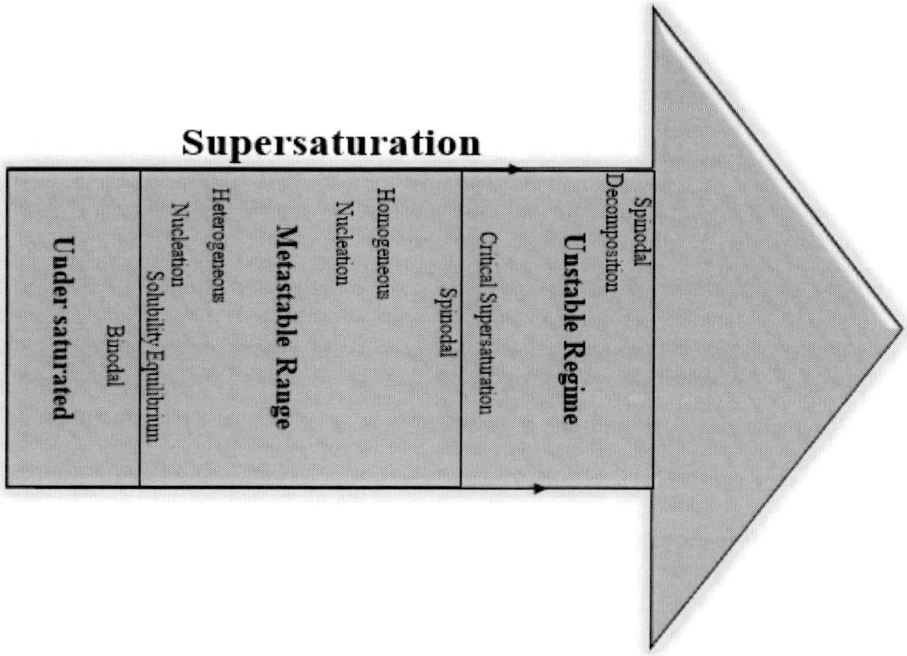

Figure 3. Effects of the phase separation mechanism on the degree of supersaturation.

$$S = \frac{C}{C_{eq}}$$ (Equation 12)

As seen in the equation describing primary nucleation rate (J_0):

$$J_0 \propto \exp\left(-\frac{16\pi\gamma^3 V_M^2 N_A}{3(RT)^3[\ln(1+S)]^2}\right)$$ (Equation 13)

where γ is the interfacial tension, V_M is the molar volume, N_A is Avogadro's number, R is the ideal gas law constant, and T is the temperature. According to equation 12, nucleation rates increase as the degree of supersaturation increases [1, 15]. The nucleation rate is, according to the classical theory of homogeneous nucleation given by equation 14:

$$J_0 = N_0 \times \upsilon \times \exp\left(-\frac{16\pi\gamma^3 V_m^2}{3k^3T^3(LnS)^2}\right)$$ (Equation 14)

where N_0 is the number of molecules of solute per unit volume, υ is the molecular transport to the solid-liquid interface frequency, γ is the interfacial tension at the solid-liquid interface, V_m is the volume of a solute molecule, k is the Boltzmann constant, and T is the temperature. The frequency of molecular transport can be approximated by:

$$\upsilon \approx \frac{kT}{3\pi\, a_0^3 \eta}$$ (Equation 15)

where η is the viscosity of the surrounding solution and a_0 is the mean effective diameter of the diffusing species. According to the classical theory, the rate of homogeneous nucleation depends on the interfacial tension at the particle-solution interface and is associated to the temperature, reversely. Furthermore, classical nucleation theory states that a rapid and high supersaturation increases the nucleation rate and consequently is the main driving force for precipitation. Up to a critical solute concentration, higher supersaturation causes a decrease in particle size. However, at higher solute concentrations, the observed particle size growth is related to the frequency increase of collisions among particles and their agglomeration. The onset of nucleation is dependent on the energy barrier (ΔG) which has to be overcome to form nuclei:

$$\Delta G = \frac{16\pi\, \sigma^3 V_s^3}{3k^2 T^2 (LnS)^2}$$ (Equation 16)

Increasing the supersaturation and lowering the surface tension can decrease this critical energy barrier, leading to faster nucleation and the creation of more and smaller particles [10, 16, 17].

The induction time, i.e., the time period between the initial supersaturation and the detection of the onset of particle formation is related to the nucleation rate. Equation 17 will relates the induction time to nucleation and growth if supersaturation during the induction time, stationary nucleation and growth kinetics, and a poly nuclear mechanism, i.e., nucleation and growth of many nuclei to a detectable size are assumed constant.

$$t_i = \left(\frac{4\alpha_v}{k_v J G^3}\right)^{1/4}$$ (Equation 17)

where t_i is the induction time, α_v is the detectable volume fraction of the newly formed solid phase that depends on the measurement device and on the substance, G is the overall growth rate in terms of diameter of a volume equivalent sphere and k_v is the corresponding volume shape factor, i.e., $k_v = \pi/6$ [14].

However, supersaturation decreases as particles grow by condensation, due to a reduction in the solute mass in solution. Therefore, condensation competes with nucleation. Moreover, coagulation competes with condensation by reducing the total number of particles. Then, final particle properties, including size distribution and morphology, are considerably affected by the processing and formulation parameters that influence nucleation and growth rates (e.g., solvent choice, stabilizer selection, and mixing rates). Therefore, faster nucleation rates, relative to growth, favor the production of uniformly smaller particles [1, 7, 10].

Model Population Balance Approach (PBA)

The model was based on homogeneous nucleation, as this phase separation mechanism is known to mostly occur in systems involving high supersaturations. Current state-of-the-art modeling of polydisperse particulate processes is based on the PBA. Consequently, the total number and mass concentrations of the particles are computed according to equation 18 and Equation 19, a corresponding mean particle size is calculated according to equation 20.

$$\frac{dN}{dt} = J$$ (Equation 18)

$$\frac{dm}{dt} = J.\frac{\pi}{6}.\rho_p.x_c^3 + N.\frac{\pi}{2}.\rho_p.x^2.G_{lin}$$ (Equation 19)

$$x = \sqrt[3]{\frac{m}{N}.\frac{6}{\pi\rho_p}}$$ (Equation 20)

where N is the total number concentration of the particles, t is the time, J is the rate of homogeneous nucleation, m is the total particulate mass concentration, ρ_p is the density of the precipitate, x_c is the critical nucleus size, x is the volume equivalent particle diameter and G_{lin} in the linear growth rate.

As growth of the gained particles was estimated to be diffusion limited, the growth rate is particle size dependent and can be written as shown in equation 21.

$$G_{lin} = \frac{dx}{dt} = 2\frac{Sh.D_{AB}.\sqrt{K_{SP}}.M}{\rho_c}.\frac{S-1}{x}$$ (Equation 21)

where Sh is the Sherwood number, KSP is the solubility product ρ_c is the density of the continuous phase and S is the supersaturation. In the mixing process described here, two phases are combined in a T-shaped mixing chamber. The global energy dissipation of the mixing process can be calculated according to equation 22, where ε is the rate of energy dissipation in the turbulent flow per unit mass of the fluid (specific power input), p is the pressure, Q is the flow rate and V_{DIS} is the effective volume in which the main turbulent energy dissipation occurs.

$$\varepsilon = \frac{\Delta p.Q}{\rho_c V_{DIS}}$$ (Equation 22)

Therefore, for Schmidt-numbers less than 4000, the characteristic micro mixing time is estimated as a function of the kinematic viscosity υ and the specific power input ε, and was calculated according to equation 23.

$$\tau_{mixing} = 17.2\sqrt{\frac{\upsilon}{\varepsilon}}$$ (Equation 23)

The formed precipitate amount and size depends strongly on the degree of supersaturation. The supersaturation S is defined by the difference of the chemical potential μ of the dissolved substance compared to a saturated solution equation 24.

$$\frac{d\mu}{RT} = LnS$$ (Equation 24)

Whether phase separation of small molecules occurs due to nucleation or spinodal decomposition. By homogenous nucleation assumption, this process can be described by equation 25. Here DAB is the diffusion coefficient, K_{SP} is the solubility, S is the supersaturation, N_A is the Avogadro's number, σ is the interfacial energy, k is the Boltzmann constant, T is the temperature and V_m is the molecular volume.

$$J = 1.5D_{AB}(\sqrt{K_{SP}}SN_A)^{7/3}\sqrt{\frac{\sigma}{KT}}V_m \exp(-\frac{16\pi}{3}(\frac{\sigma}{KT})^3\frac{V_m^2}{(\ln S)^2})$$ (Equation 25)

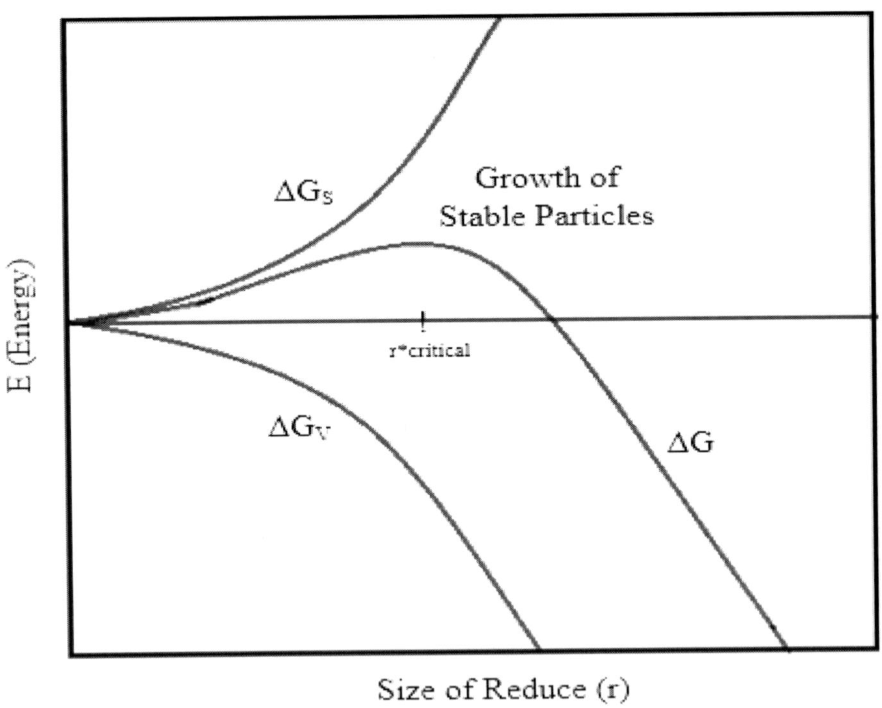

Figure 4. Energy diagram to explain the nucleation process [18].

For nucleation based decomposition an energy barrier has to be overcome in order to form thermodynamically stable nuclei. The free energy of a spherical nucleus with radius r can be described in respect of the single-phase state to a first approximation as in equation (25) with μ* being the difference in chemical potential between a solid and the corresponding dissolved substance [7, 18].

$$\Delta G = \Delta G_s + \Delta G_v = 4\pi r^2 \gamma + \frac{4}{3}\pi r^3 \Delta g_v = 4\pi r^2 \gamma + \frac{4}{3}\pi r^3 \rho_c d\mu^*$$

(Equation 26)

Since the two terms on the right of the equation 26 have differing signs, the ΔG as a function of r passes through a maximum (Figure 4) [18].

It is obvious that both the critical radius (r*) and the critical activation free energy of nucleation (ΔG*) are contrariwise proportional to the degree of supersaturation. Hence, higher degrees of supersaturation, reduces ΔG*, which in turn increases the nucleation rate (J). The other factors, which control the degree of supersaturation as discussed before, indirectly affect the nucleation rate [19]. The critical nucleus radius r*, corresponding to the maximum of the free energy, is given by:

$$r^* = \frac{-2\gamma}{\Delta g_v}$$

(Equation 27)

Δg_v is the difference in free energy per unit volume between the two phases [3, 18]. Only for particles of a certain critical size the negative volume part of the function dominates the thermodynamically unfavorable positive surface part, the nucleus becomes stable and grows by combination of further molecules to the existing particle. The critical nucleus size can be calculated by differentiating the free energy with respect to the nucleus radius r to determine the maximum of the free energy, where the critical nucleus size is given by equation 28.

$$x_c = \frac{4\gamma V_m}{kT\, LnS}$$

(Equation 28)

Besides supersaturation, the nucleation rate depends mainly on the interfacial energy σ, which controls the interfacial actions of a particle against the surrounding fluid. The interfacial energy is also altered significantly, when the composition of the continuous phase changes within the precipitation process.

The diffusion coefficient D_{AB} was estimated based on the Stokes-Einstein equation (equation 29), the molecular volume V_m was determined based on the molecular weight and the density of the drug (equation 30) [7].

$$D_{AB} = \frac{kT}{6\pi \eta_c r}$$

(Equation 29)

$$V_m = \frac{M}{\rho_p N_A} \quad \text{(Equation 30)}$$

The Forces between the Particles in Nanodispersion Systems

The control and stability of nanodispersions has been the subject of extensive research because of its great importance in many industries such as pharmaceutical, ceramic, paints and pigments [20]. The stability of nanodispersions depends on the size of nanoparticles, the collision frequency between nanoparticles and the sticking efficiency of these collisions [21]. The colloid interactions are very short range since the particles have to approach very close to each other, their charged double layers overlap and repulsion takes place [20, 22]. Interactions between colliding particles must be such that permanent contacts can be formed. Particles which repel each other are said to be stable, since they do not form aggregates. The most important forces that determine the stability of the colloidal dispersions are attractive and repulsive forces [20, 23]. The importance of the double layer and van der Waals interactions has been well demonstrated by the classical DLVO theory (Verwey and Overbeek, 1948). DLVO theory accounts for both electrostatic repulsion and van der Waals attraction between two approaching particles as a function of interparticle distance [24]. The repulsive forces must be prevailed for maintaining the stability of the colloidal system. Stabilization of nanocrystals in an aqueous solution is based on two basic mechanisms: steric stabilization and charge or electrostatic stabilization [20, 22, 23]. In the system, electrostatic stabilization takes place due to the distribution of charged species and its effect on particle interaction [20]. Electrostatic interactions occur between charged surfaces which are diffuse clouds of counter and co-ions surrounding charged surfaces in aqueous solution [21]. The charge around the particle is arranged into a double layer in which each layer possesses an equal charge. This mechanism relies on the separation of ionic charges; therefore, it is mainly relevant in systems of high polarity like aqueous paints. In electrostatic stabilization the nanodispersions can be stabilized or flocculated by changing the concentration of ions in the system [20]. The second interaction component accounts for Van der Waals interactions, which are constant and fixed by the solid bulk composition [21]. The opposing effects of attractive dispersion and repulsive steric forces occur in most asymmetric systems when the alignment of maximum interaction energy is also the one where the molecules can come together, which is the orientation that wins out. Steric Interaction Adsorbed layers, mainly of polymers, can play a very important role in aggregation and deposition phenomena. In some cases, small amounts of adsorbed polymer can increase flocculation by a 'bridging' mechanism. However, in most cases, polymers are added to the system adsorbing onto the particle surface which prevents the particle surfaces coming into close contact due to steric repulsion [20, 22, 25].

The nanoparticles stabilization may be carried out with electrostatic, steric and electrosteric effects, which would be improved by using pH or ionic strength, surfactants and polyelectrolytes, respectively [26]. Electrosteric stabilization can also be provided by the use of a combination of two different stabilizers. Combination of more than one stabilizer has sometimes been preferred for enhanced long-term stability. The stabilizer effect of ionic surfactants is due to the formation of surfaces with a charge sufficient for stabilization [23]. In nanodispersion systems polymers can give greatly enhanced stability, by an effect which is usually known as steric stabilization [22]. The net magnitude of the ξ- potential gives an indication of the potential stability of the dispersion system. Therefore, the particles with higher ξ- potential values, either positive or negative, will be more stable. Physically stable nanodispersions stabilized only by electrostatic repulsion should have a minimum zeta potential of ±30 mV in order for the dispersion to resist flocculation [27].

CFD Simulation

As stated before, solvent-displacement or nano-precipitation process a novel technique to produce functional nanoparticles stabilized by amphiphilic compounds. Mixing in different microreactors, such as confined impinging jets (CIJ) and multi-inlet vortex reactors (MIVR), has been investigated both experimentally and through simulations. To investigate the mixing process in the MIVR, a computational fluid dynamics (CFD) model was developed to predict and compare with the experimental data. To describe the particle formation and size distribution as functions of mixing time and physical properties of inlet streams in the NP process, a competitive aggregation model has been developed. Using this model, knowledge of particle properties can be obtained by solving a bivariate population balance equation (PBE) [28, 29].

The deviation from the well-mixed behavior in a crystallizer is primarily caused by the hydrodynamic conditions, which lead to a supersaturation profile over the crystallizer. Recently, CFD techniques have been developed to a state that allows the complex flow situation to be modeled with a reasonable amount of detail and accuracy. A hybrid approach integrating CFD techniques and process simulation tools can, therefore, be of great help in grasping the complex interactions between hydrodynamics and the other physical and chemical phenomena [30]. A coupled CFD micromixing population balance algorithm can used to model the crystallization process in an impinging jet crystallizer. The full crystal size distribution for the anti-solvent crystallization of lovastatin in the confined impinging jet crystallizer was simulated by coupling the macromixing and micromixing models with the solution of the population balance equation as described by Woo et al. and the effects of jet velocity can numerically investigated. The development of the CFD population balance coupled algorithm is presented in detail by Woo et al. [31], which applied the integrated model to simulate semi batch anti-solvent crystallization in a stirred tank [29, 31].

The macromixing was modeled by the Reynolds-averaged Navier Stokes (RANS) and transport equations and the k-ε turbulence model with enhanced wall treatment due to the small volume of the mixing chamber and the confinement of the impinging region by the chamber walls [31]. The micromixing (mixing at a molecular level) was also modeled by a three-environment ($N = 3$) presumed probability density function (PDF) micromixing model, which approximates the fluctuations of the species concentrations on the subgrid scale. One inlet stream (saturated solution) is represented by Environment 1, and the other inlet stream (anti-solvent) is represented by Environment 2. Environments 1 and 2 transfer matter by engulfment into Environment 3, where crystallization occurs in a supersaturated solution [31]. To compute the crystal size distribution in the impinging jet crystallizer, the transport equations in the CFD-micromixing model are coupled to a spatially varying PBE that this is discussed in detail by Woo et al. [32]. The spatially varying population balance equation, discretized along the crystal growth axis using a high-resolution finite-volume method, is solved as a set of scalar transport equations in the CFD solver. The macromixing and micromixing models are coupled with the population balance in a commercial CFD solver [31]. It was indicated that crystal growth for this system is surface integration limited, and the growth rate is size-independent. Secondary nucleation can be ignored due to the small solids density in impinging jets [33].

Effects of Jet Reynolds Number on particle size distribution have also been discussed. The monotonically decreasing shape of the particle size distribution computed from the simulations is very similar to that obtained in the experiments by Mahajan and Kirwan [33] for the same crystallization system for a nonsubmerged unconfined impinging jet [31]. The capability of the CFD micromixing population balance model in modeling crystallization in confined impinging jets, considering the effects of macromixing and micromixing, was demonstrated as well. The development of the crystallization process would be done in the following steps; for a given anti-solvent solvent solute crystallization system, hydrodynamics-independent crystallization kinetics would be first determined by using small amounts of the compound. With the estimated parameters for the CFD micromixing model and crystallization kinetics, the CFD micromixing population balance model is used to simulate the crystal size distribution for a wide range of operating conditions (for example, inlet velocities and inlet concentrations). The simulation results then can be used to identify the set of operating conditions that produces the crystal size distribution that meets the bioavailability requirements [29, 31]. The MIVR, the mixing is always macromixing controlled, and moderate turbulent flow will yield similar mixing results as highly turbulent flow [28].

Classification of Solvent-Displacement Technique

The developed and patented solvent-displacement technique for nanoparticle manufacture was first developed and patented by Fessi et al. [34], presents various advantages, including it does not require any extended stirring rates, sonication or high heating processes, all conditions that might damage bioactive compound structure. Additionally, are unacceptable toxic surfactant and organic solvents are generally excluded from this procedure. Nanoprecipitation often enables the production of small nanoparticles (<300 nm) with narrow unimodal distribution and a wide range of natural polymer or biopolymers can be used for long term stabilization of nanoparticles [5]. This bottom-up, very easy, efficient and cost-effective, scalable size reduction technique, forming both nanocapsules and nanospheres, is shown schematically in Figure 5 [5, 35]. Briefly, it requires two solvents that are miscible. Ideally, both the polymer and the drug must dissolve in the first one (the solvent), but not in the second system (the anti-solvent) [5]. In precipitation techniques, the poorly water-soluble drug is typically dissolved in a solvent and precipitation of the drug is initiated by a reduction in solvent power, by either addition of an anti-solvent or solvent evaporation. A reduction in solvent powder leads to supersaturation of the drug and drives nucleation of drug particles. Once nucleation occurs, the particles grow by condensation, in which dissolved drug molecules diffuse to the particle surface and integrate into a solid particle, and/or by coagulation, where multiple particles collide and aggregate to form larger particles. The rapid nanoparticle formation is governed by the so-called Marangoni effect, which is due to interfacial turbulences that take place at the interface of the solvent and the non-solvent and result from complex and cumulated phenomena such as flow, diffusion and surface tension variations [5, 36]. Stabilizers may be added to the system to arrest particle growth [37]. In such processes phase separation typically occurs within a time frame of μs to ms, so that with a very high likelihood amorphous precipitates can be obtained [7]. Anti-solvent precipitation has been widely investigated in the pharmaceutical industry [38, 39]. However, although technologically closely related to the widely applied crystallization techniques [7].

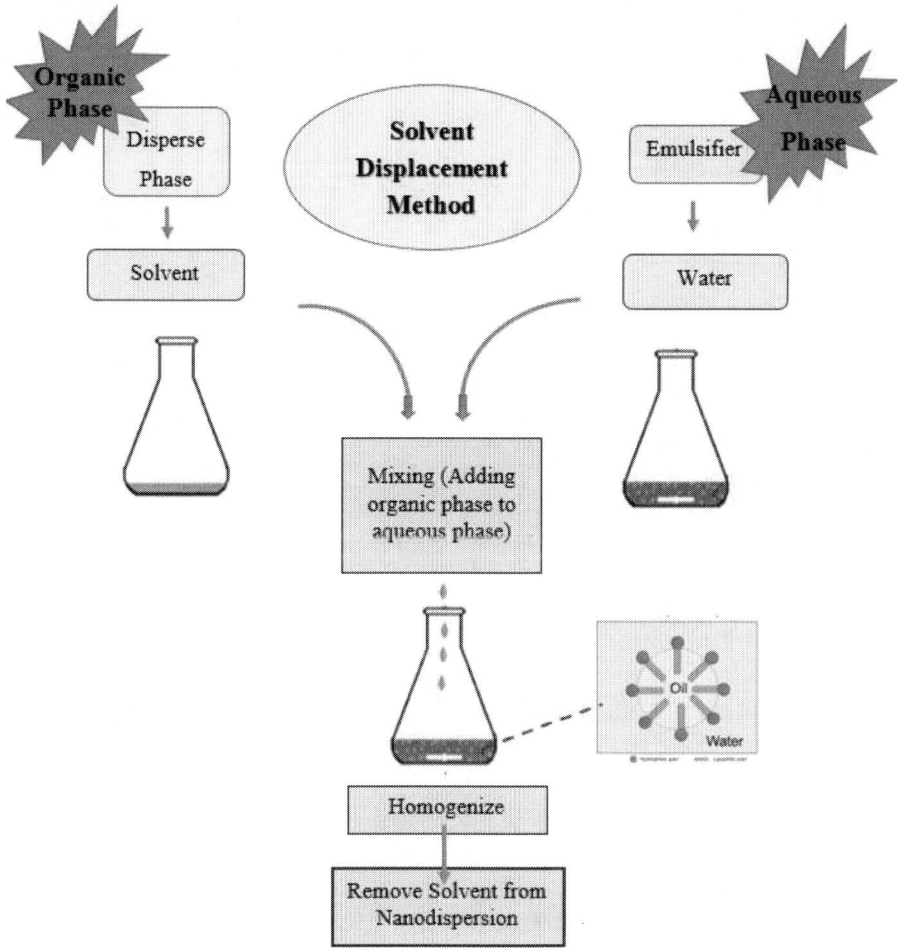

Figure 5. solvent-displacement process steps.

Liquid Anti-Solvent (LAS) Precipitation

The liquid anti-solvent (LAS) precipitation process offers more expedient procedure at ambient temperatures and atmospheric pressure with no requirement of expensive equipment. It is easily accessible as compared to other bottom-up methods. This method can be analyzed in various aspects such as, nanoparticles formation, process intensification, and stabilization of aqueous suspensions of nanoparticles [10, 40].

The main driving force of the formation of particles during anti-solvent precipitation is the imbalance of molecular interact ions between solute, solvent, and anti-solvent due to the super saturation. Certain solvent/anti-solvent combinations are determined based on the nature of the involved polymers or compounds. At this rate the solute -solute interact ions are strong enough to overcome other interactions and entropy of mixing effects, resulting in precipitation. After the particles are formed it is important that there is a strong repulsion between them to avoid them from aggregating. Different production stages can be distinguished in the anti-solvent precipitation process: generation of super saturation; nucleation; crystal growth; and, coagulation, as stated before [10].

Anti-Solvent Precipitation Using Organic Solvents

The anti-solvent precipitation (AP) process, in which organic solvents make up the solvent phase, is one of the most common bottom-up approaches for particle formation [41]. Due to the organic compounds burden large amount of solubilities compared to supercritical fluids, the larger process yields are generally obtained from AP operations. In AP, the poorly water-soluble bioactive compound is first dissolved in an organic solvent and then the bioactive compound containing solution is mixed with an anti-solvent, often water. As the two phases mix, the solubility decreases, resulting in supersaturation, which efforts nucleation. An organic solvent that is miscible with the anti-solvent is selected to facilitate mixing between the two phases. Diffusion of the organic solvent into the aqueous phase tends to spread the nuclei apart, reducing coagulation rates, in which droplet shrinkage during solvent evaporation tends to promote coagulation of nuclei. Stabilizers may be added to the solvent or anti-solvent phases to moderate particle growth by condensation and coagulation. The hydrophilic segments of the stabilizer favorably extend toward the aqueous environment and, thus, the stabilizer adsorbs at the interface and not within the interior of the particle. Passivation of the particle surface by the stabilizers prevents particle growth. The selective orientation of the stabilizer at the particle surface facilitates the production of stable, high-potency nanoparticles with a minimum stabilizer ratio. An illustration of the driving mechanism for the AP process is shown in Figure 6 [37].

Controlled Precipitation Process

To minimize particle growth after precipitation, the controlled precipitation (CP) process includes a semi-continuous solvent elimination step, such as vacuum distillation, after the solvent and anti-solvent are mixed together. By including the solvent removal step in-line with the process to remove excess solvent immediately after precipitation instead of in a separate step, substantial levels of particle growth may be minimized [37].

In CP, solvent removal is typically performed within a wiped film evaporator to maximize the presented surface area over which solvent evaporation may occur, as well as to reduce any foaming that might have occurred during processing, given the high amounts of emulsifier required for stabilization in some cases. Rapid removal of a significant portion of the solvent markedly reduces the solubility of the lipoid compound in the mixed solvent.

Thus, particle growth by condensation and Ostwald ripening can be minimized considerably. Another important aspect of the CP process is its high propensity to produce amorphous particles, due to the rapid nucleation and stabilization rates generated during the precipitation process [37, 42].

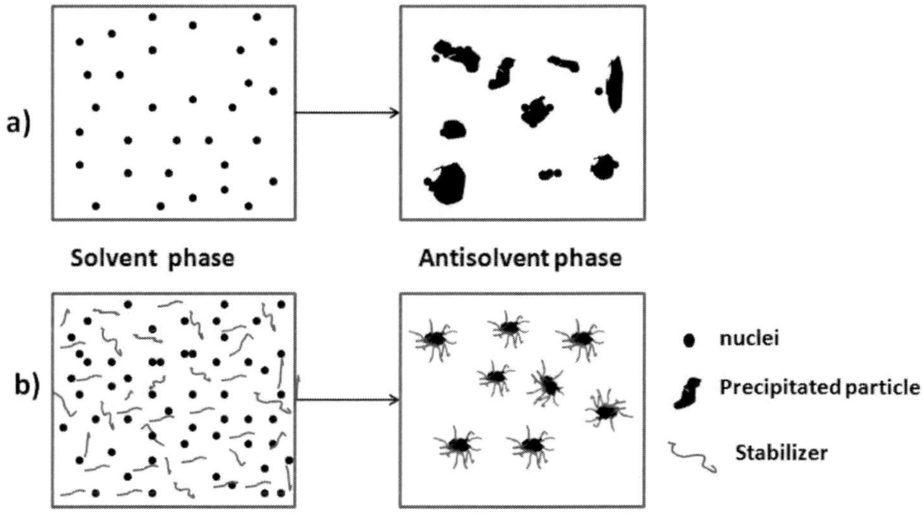

Figure 6. Anti-solvent precipitation (AP) of organic nanoparticles in the (a) absence and (b) presence of amphiphilic stabilizers.

Ultrasound Assisted Anti-Solvent

Ultrasound has received much attention to be used as an effective measure to influence the nucleation in crystallization processes. The use of ultrasonic waves in crystallization has been increased in order to; (i) rapid and uniform nucleation throughout the sonicated volume leads to smaller and uniform-sized particles, and (ii) reduction of agglomeration of particles and controlling the number of nuclei [43].

The mean particle size and its distribution can be effectively controlled by adjusting ultrasound parameters such as the power intensity and ultrasonic time during crystallization. Sono-crystallization can also be used for the high energy materials which are sensitive to friction and impact during size reduction by mechanical means [44]. The influence of an ultrasonic wave on particle size can be studied by changing three basic parameters: ultrasound addition time, ultrasonic power intensity, and the time at which the sonication is applied. When ultrasound waves proliferate through a liquid medium, its power will initiate an important phenomenon known as cavitation. The formations of bubbles are occurring during the negative pressure period of the sound wave during cavitation. These bubbles will rise up to their resonance size and then they collapse, causing a localized hot spot with a high temperature and pressure including the release of powerful shock waves. The power of ultrasound and cavitation phenomena will initiate the nucleation and thereby crystal growth in a crystallization process. The use of ultrasound may also influence the solubility and thereby the supersaturation solubility. It can also alter the crystal pattern as the ultrasound can increase or decrease the growth rate of certain crystal faces. Localized hot spots may influence the crystal lattice and have some effect on the crystal habit change due to abrasion [45].

The most important effect of the presence of an ultrasonic wave during crystallization is that it reduces both the nucleation induction time and the metastable zone width. Hence, the ultrasound may control the rate of nucleation and crystal growth and thereby affect the resulting size of each particle [45].

Emulsion-Solvent Evaporation Based Precipitation

This is one of the most frequently used methods for the preparation of nanoparticles. Emulsification-solvent evaporation involves two steps. The first step requires emulsification of the functional lipid bioactive compound solution into an aqueous phase. During the second step solvent is evaporated, inducing bioactive compound precipitation as nanoparticles. The nanoparticles are collected by ultracentrifugation and washed with distilled water to remove stabilizer residue or any free compounds.

High-pressure emulsification and solvent evaporation method is also known as a rectified techniques. This method involves preparation of a emulsion which is then subjected to homogenization under high pressure followed by overall stirring to remove organic solvent. The size can be controlled by adjusting the stirring rate, type and amount of dispersing agent, viscosity of organic and aqueous phases and temperature. However this method can be applied to liposoluble compound and limitation are enforced by the scale up issue [46].

Double Emulsion and Evaporation

The emulsion and evaporation method suffer from the limitation of low entrapment efficiency of hydrophilic drugs. Therefore to encapsulate hydrophilic drug the double emulsion technique is employed, which involves the addition of aqueous drug solutions to organic polymer solution under vigorous stirring to form w/o emulsions. This w/o emulsion is added into second aqueous phase with continuous stirring to form the w/o/w emulsion. The emulsion then subjected to solvent elimination by evaporation and the nanoparticles can be produced after this step. In this method the amount of hydrophilic compound to be incorporated, the concentration of active compound, stabilizer used, the volume of aqueous phase are the variables that affect the characterization of nanoparticles [46-50].

Supercritical Fluid Based Precipitation

Supercritical fluids (SCF) exhibit properties intermediate between those of liquids and gases, such as low viscosity, low density, high solvating power, high diffusivities, and high mass transfer rates above the critical point. A number of compounds can be brought to a supercritical state, such as carbon dioxide, water, propane, nitrogen, etc. Supercritical fluid technology is employed through some methods including: rapid expansion from supercritical solution, gas anti-solvent, supercritical anti-solvent precipitation, aerosol solvent extraction, and precipitation with a com-pressed fluid anti-solvent [51]. SCF processes represent alternative techniques to conventional routes of particle generation (micronization and co-precipitation/coating). Indeed, these processes allow the formation of smaller particles with a narrower particle size distribution as well as an accurate control over polymorphic purity and particle morphology [52].

Supercritical fluids are used for encapsulation of thermally sensitive compounds in a process similar to spray drying. In this technique, the bioactive compound and the lipophilic compound are solubilized in a supercritical fluid and the solution is expanded through a nozzle. The supercritical fluid is evaporated in the spraying process, and solute particles eventually precipitate. This technique has been widely used for low critical temperatures and minimum use of organic solvent [53]. Through the mild temperature conditions that could be applied especially when CO2 is the SCF (Tc = 304.21 K), thermo-sensitive drug molecules can be manipulated without degrading their properties. These methods use SCFs, generally supercritical CO2 (scCO2), either as a solvent in rapid expansion from supercritical solution processes (RESS), as an anti-solvent in supercritical anti-solvent (SAS) processes or as a solute in particles from gas-saturated solution processes (PGSS). The most studied SCF micronization technique is the SAS process. This technique is used to precipitate solutes (generally polar compounds) with low solubility in supercritical fluid by taking benefit of the miscibility of scCO2 with several organic solvents. Moreover SAS process using scCO2 allows us to obtain controlled particle size and distribution that are not achievable by conventional methods [52].

Compressed fluid and supercritical fluid (SCF) anti-solvent precipitation processes offer several advantages over more conventional liquid anti-solvent processes. The anti-solvent may be completely removed via pressure reduction to the gaseous phase, resulting in improved product purities and reduced environmental and toxicity concerns [52].

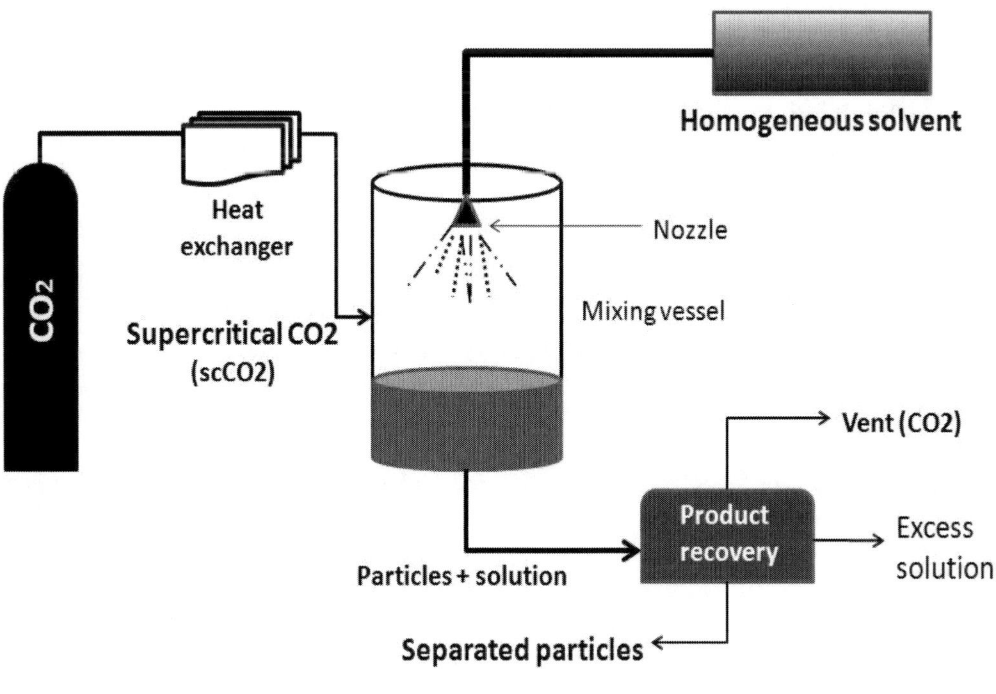

Figure 7. Supercritical anti-solvent technique.

Table 1. Common supercritical fluid with their critical constants [54]

SCF	TC (centigrade)	PC (bar)
Carbone dioxide	31.1	73.7
Propane	96.8	43
Ammonia	132.4	112.7
Water	374	220.5
Ethane	32.3	48.8
Chlorotrifluoromethane	28.9	39.2
Trifluoromethane (fluoroform)	25.9	47.5
Dinitrogen monoxide	36.5	72.6
Sulfurhexafluoride	45.5	37.6
Chlorodifluoromethane (HCFC 22; R22)	96.4	49.1
Trichlorofluoromethane (CFC 11; R11)	198	44.1

An important feature of SCFs is that their densities can change significantly with small changes in pressure. These density changes give rise to variations in diffusivity and viscosity, as well as the solubility of other solvents and small solutes. Typical diffusivities of SCFs are on the order (~100 times greater than that for liquids) and their viscosities are on the order of (~100 times lower than that for liquids). These favorable mass transfer properties facilitate rapid diffusion of the SCF anti-solvent into a liquid solvent, which enables rapid supersaturation and nucleation, thus favoring the production of small particles [37]. The common used SCFs are listed with their critical pressure and temperature in Table 1 [54]. Besides these fluids, carbon dioxide (CO_2) is the most prevalently used in pharmaceutical applications because it is inexpensive, nonflammable, and nontoxic.

Rapid Expansion of Supercritical Solution

The rapid expansion of supercritical solutions (RESS) process utilizes the SCF as a solvent, not an anti-solvent. The solute is dissolved directly into the SCF phase in the extraction unit. Then, the system is depressurized across a nozzle into a collection chamber at atmospheric conditions. The rapid depressurization origins evaporation of the SCF, resulting in a significant decrease in solvent power, and thus promotes rapid nucleation and precipitation of the solute. As with the other particle formation techniques discussed previously, additional excipients may be dissolved in the SCF, typically CO_2, to produce composite particles of drug and excipients (Turk 2009). The schematic of the RESS process is shown in Figure 8. Intense atomization of the drug–CO_2 stream is desirable to achieve nanoparticles from the RESS process. Therefore, depressurization of the CO_2 feed stream from the nozzle is designed to be extremely rapid, with typical CO_2 flow rates exiting the nozzle at the speed of sound, creating super-saturation levels on the order of 105 –106 within a time frame 10 –6 –10 –4s. The intense turbulence generated by rapid depressurization of CO2 distributes the newly generated supersaturation regions almost instantaneously and homogenously throughout the fluid, which facilitates the production of small particles with narrow PSDs. This rapid dissipation of energy is highly endothermic, and thus the nozzle is generally heated to prevent freezing of CO_2 during atomization, which can cause clogging.

In traditional RESS, the solute is dissolved in a supercritical fluid to form a solution, followed by the rapid expansion of the solution across an orifice or a capillary nozzle into ambient air. The high degree of super saturation, accompanied by the rapid pressure reduction in the expansion, results in homogenous nucleation and thereby, the formation of well-dispersed particles. Results from mechanistic studies of different model solutes for the RESS process indicate that both nanometer and micrometer-sized particles are present in the expansion jet [55].

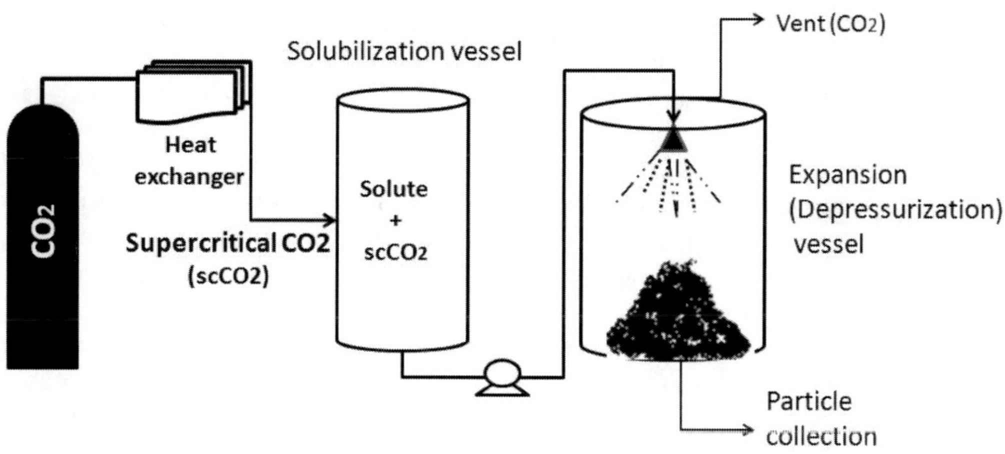

Figure 8. Rapid expansion of supercritical solution (RESS) process.

Modifi ed RESS Processes (RESOLV)

A simple, but significant modification to RESS involves expansion of the supercritical solution into a liquid solvent (RESOLV) or into a aqueous solution (RESAS) containing a stabilizer, instead of ambient air, termed as RESOLV [37, 55]. The main disadvantage of RESS process is the micro-scaled of the preliminary particle size, instead of nano-scaled. To overcome this drawback, the RESOLV method has been employed. In RESOLV the liquid solvent apparently suppresses the particle growth in the expansion jet, thus making it possible to obtain primarily nanosized particles [56].

Rapid Expansion of Supercritical Solutions with Solid Co-solvents (RESS-SC)

In the RESS-SC process, a co-solvent that solidifies upon atomization from the nozzle is used to enhance the solubility of solutes in scCO2, as well as provide a barrier for coagulation in the free jet during scCO2 expansion [57]. In contrast to RESS, where the nuclei tend to coagulate during free jet expansion, the excess amounts of solid co-solvent added during the

RESS-SC process surrounds the nuclei to create a physical barrier to reduce coagulation. The co-solvent may be removed later by lyophilization [37].

Gaseous Based Anti-Solvent Precipitation (GAS)

In GAS precipitation method, CO_2 is used as a SCF anti-solvent, which is accompanied by an organic solution including active pharmaceutical ingredients (API). GAS process is usually applied in high CO_2 solubility pressure (5-8 MPa) in most organic solvents [58]. As the CO_2 dissolves into the solute-rich liquid phase, the solvent strength decreases. Subsequently, the occurred decrease in the active compound solubility in the solvent, leads to nucleation and then precipitation followed by supersaturation. In some cases, additional excipients may also be dissolved in the lipid based organic compound solution to precipitate it within an excipient matrix. In order to induce rapid nucleation, which favors the production of small particles, CO_2 must be readily soluble in the organic solvent and the active lipid based compound must have low solubility in CO_2. Under optimal operating conditions, CO_2's high solubility and favorable transport properties in the organic solvent facilitate homogenous supersaturation conditions more rapidly than can be achieved using liquid anti-solvents. When precipitation is complete, the CO_2 –organic solvent solution is flushed from the system and the precipitate, i.e., the active compound powder, remains in the precipitation vessel. It may then be washed with fresh CO_2 to remove excess organic solvent. The primary problem of the GAS precipitation process is the difficulty in harvesting the precipitate particles from the organic solvent solution though minimizing particle growth and agglomeration. Furthermore, in cases where elevated temperatures are need to sufficiently expand the SCF into the organic solvent, thermal degradation of the API may occur [37]. Because GAS precipitation is driven by the anti-solvent abilities of CO_2 in the organic solution, appropriate processing conditions may be selected based on optimizing thermodynamic criteria, specifically by understanding the volumetric expansion of the organic solvent due to CO_2 solubilization, and thus solubility of the solute in the solvent–CO_2 mixture [37].

Flash Nanoprecipitation Process

Flash nanoprecipitation (FN) has been developed to facilitate the production of amorphous nanoparticles for enhanced dissolution of poorly water-soluble drugs, aiming to minimize mixing times between the solvent and anti-solvent, using a custom-designed confined impinging jet (CIJ) mixer. In a CIJ mixer, a solvent stream and an anti-solvent stream are introduced into a mixing chamber at turbulent jet velocities, in such a manner that the two streams are collinear and thus collide with each other. Mixing within impinging jets produces a region of high-energy dissipation, as the kinetic energy of the jet streams is converted to turbulent motion through collision and redirection of fluid flow within a confined volume. These high-energy dissipation regions rapidly reduce the scale of segregation between the two fluid streams, thus facilitating rapid nucleation. The mixing chambers in the FN process must be large enough for the high-energy dissipation regions to form, but limited in volume to avoid significant bypassing of any fluid from intense [37].

Flash NanoPrecipitation has been used for the formation of paclitaxel-vitamin E succinate conjugate nanoparticles using methoxy poly (ethylene glycol)-b-poly (ε-caprolactone) (mPEG5-PCL7) block copolymer [59].

Precipitation in Acid

Scientists have recently presented a modified approach to AP, in which a poorly water-soluble drug compound is stabilized by an ionic polymer via precipitation in an acid bath [60]. The bioactive compound and ionic polymer are dissolved in an organic solution, which is then added to a cooled acid bath to initiate nucleation of the drug particles. The ionic nature of the stabilizer is critical for effective "micro-precipitation" of "nanosized" domains, which are claimed to be molecularly dispersed throughout a polymer matrix. The resultant powder has been termed a "microprecipitated bulk powder" (MBP). Amorphous morphologies have also been formed under sufficiently rapid stabilization conditions. Ionic polymers with a MW of at least 80,000 Da and a glass transition temperature, $T_g > 50°C$ have been found to promote efficient stabilization of submicron, amorphous drug domains by this precipitation technique. The acidic aqueous bath is likely needed to elicit the desired charge on the polymer to promote a strong interaction between the polymer and the newly formed drug surfaces. This technology has demonstrated scalability and is currently being used for the production of several Biopharmaceutical Classification System Class II (low-solubility, high-permeability) and Class IV (low-solubility, low-permeability) compounds in phase I–II clinical trials [37, 61].

Effects of Main Processing on Characteristics of Produced Nanodispersions

The controlling of the size distribution and stability of produced nanodispersions via solvent-displacement procedure is very important for most applications. In most uses production of the smallest monodispersed fine nanoparticles with highest long term stability are preferred. This goal can be achieved by altering main process and formulation parameters, which are going to be discussed.

Effect of Addition Rate

The previous researches have been shown that the crystallization at constant temperature depends on the water-addition rate and the initial concentration of the solution. The crystallization at constant temperature depends on the water-addition rate and the initial concentration of the solution. Since at low initial concentrations, the crystallization process is independent of the water-addition rate, in higher initial concentrations, the crystallization form considerably can be influenced by addition rate, especially at less rates of water-addition [62]. The addition order (solvent to anti-solvent or anti-solvent to solvent) has also a major impact on the particle properties. The addition rate and order determine the degree of supersaturation, and therefore strongly influence the nucleation rate and particle growth kinetics [10].

It has been reported that increasing the flow rate of an organic phase (solvent with dissolved compound) into an anti-solvent significantly decreased the mean size of the produced nanoparticles, which was attributed to more efficient mixing step. Increasing the flow rate caused an increase in all jet velocity, Reynolds number and shear forces, which improve the extent of mixing per unit time and leads to a considerable reduction in particle size [63]. Lander et al. [64] also found a substantial effect of the addition rate on the polydispersity index of the particles, but not on the particle size.

In general, the metastable zone widened with increasing addition rate, with the effect most pronounced when the anti-solvent was added close to the impeller. At this location, an increase in agitation intensity resulted in a narrower metastable zone for all addition rates. For an addition location close to the vessel wall, the metastable zone was narrower and the impact of addition rate and agitation were less pronounced. Substantial variation in the measured metastable zone width was also observed, with nucleation occasionally occurring at bulk concentrations less than the saturation level. It is proposed that the metastable zone width is influenced by the differing degrees of anti-solvent incorporation at each addition location. Close to the impeller anti-solvent is rapidly incorporated leading to consistent results, but, close to the vessel wall, incorporation is hindered by unfavorable mixing conditions leading to premature nucleation and more variability [65].

It was also resulted that the particle size would be larger if the anti-solvent was added to solvent rather than addition of solvent to anti-solvent [10].

Anti-Solvent to Solvent Ratio

The comparative volumes of anti-solvent and solvent used to prepare nanodispersions also have an important effect on nanoparticle characteristics. Previous studies have shown that the particle size will be reduced by increasing the volume of used anti-solvent to solvent ratio [49, 63]. The instantaneous increase of supersaturation due to the more rapid reduction in solvent concentration and decrease in particle growth at lower solute concentrations are some explanation for this observation. After formation of the nuclei, particle growth occurs which is partially hampered at higher anti-solvent volumes as the diffusion distance for the growing species increases and the process becomes diffusion limited [10, 63].

Effects of Mixing Speed

The mechanical stirring causes an even degree of supersaturation through the system. For mixing the solvent and anti-solvent phases, three different turbulent mixing scales as macro-, meso- and micro-mixing, can be considered. The total mixing time (τmix) is the needed time for all mentioned mixing scales.

$$\tau_{mix} = \tau_{macro} + \tau_{meso} + \tau_{micro} \qquad \text{(Equation 31)}$$

It has claimed that the mixing time for meso and micro scales are more substantial. The term meso-mixing concentrated on the coarse-scale turbulent mixing of the solvent and anti-solvent in the case of anti-solvent precipitation, which is influenced by the solvent feed rate, stirring speed, and the turbulent diffusivity. The macro- and meso- scales mixing create the environment for the micro-mixing scale. Micro-mixing is mixing on the smallest scale of motion and covers molecular diffusion. Micro-mixing directly controls the local distribution of supersaturation in a system and consequently, the nucleation step, while meso- and macro-mixing, in contrast, affects the particle growth step [10, 63].

During the creation of nanoparticles, the mixing time should be noticeably shorter than the sum of the nucleation and particle growth time, known as induction or precipitation time (τ_{ind}). Therefore, the dimensionless Damkohler number which is the ratio between the mixing and precipitation times should be smaller than 1.0. If the Damkohler number is greater than 1.0, the metastable zone will be crossed slowly. Thus particle growth will be promoted instead of nucleation and larger crystals will be formed. Effective micro-mixing leads to an effective enhancement of the mass transfer and the rate of diffusion between the phases, which cause rapid formation of homogeneous supersaturation and therefore rapid nucleation and smaller particles production. The position at which the anti-solvent is fed into the solute solution is another main parameter for the nucleation rate and particle size growth. Increasing the mixing rate or decreasing τ_{mix} can be achieved by using appropriate mixing devices, while increasing the precipitation time can only be achieved using stabilizers. Elevated mixing rate disrupts agglomerates and produces smaller mean particle size as well [10, 37, 63, 66].

In the last decade, ultrasound has received much attention and has been used as an effective measure to reduce the induction time and speed up the nucleation rate, by changing the solubility, diffusion coefficient, micro-mixing and meta stable zone size [10, 67]. The ultrasound increases the frequency of the particle collisions, but at the same time reduces the contact time between the crystals, as well. Thus it affects the particle agglomeration twofold. Moreover, ultrasound also causes particle size reduction by abrasion. The frequency, intensity, duration and horn tip diameter of the ultrasound are some of the main parameters in this regard. Furthermore, the soaking depth and volume of solution can also play an important role in precipitation steps [10, 67].

Sheng et al. [68] investigated effect of sonication power on the particle size. Presence of ultrasound during precipitation effectively reduced the particle size. Further increase in sonication power resulted in a gradual size reduction. This is ascribed to the effective acceleration of mass transfer by high ultrasound power [69]. Also showed with one minute of sonication, the size of precipitated β carotene particles was decreased rapidly. This indicates that ultrasound would enhance the mixing of fluid and decrease the mixing time through increased energy dissipation rate, thus decrease the particle size [68].

Active Compound Concentration

The concentration of the active compound in organic phase has also important effect on the particle formation process through two main mechanisms; First, a greater solute concentration increases the degree of supersaturation, leading to the formation of more and smaller nuclei, however, simultaneously nuclei growth will be promoted due to increasing the frequency of particle collisions and decreasing the diffusion length. Second, increasing solute concentration can also increase the viscosity of system, hindering the diffusion between solvent and anti-solvent, which lead to heterogeneous supersaturation and slowing down the nucleation rates. On the other hand, increasing the viscosity will also decrease the number of collisions, the mass transfer through diffusion, and hinder the particle growth [10, 63, 70].

Temperature

Temperature affects the precipitation process by changing the equilibrium saturation and supersaturation concentration, the diffusion rate and the viscosity of the system. All these factors will influence the particle size by controlling nucleation rate, particle growth, coagulation and agglomeration. In general, higher temperatures caused the production of polydispersed larger nanoparticles. Moreover, at higher temperatures, increasing the solubility of the compound, slowing down the nucleation rate, leads to the formation of larger particles. The raised temperatures also encourage diffusion and increase the growth kinetics at the particle surface. A decrease in anti-solvent temperature, depending on the anti-solvent used, will generally lead to an increased viscosity, which can decrease the molecular transport rate and reduce the particle collision frequency and reduction of particle growth by coagulation as well as agglomeration. Totally, the used emulsifiers and biopolymers can be structurally denatured by heating, which may also change the particle formation and stability [10].

Selection of an Appropriate (Anti-)Solvent

The nature of used anti-solvent is the most important to control the precipitation process. A prerequisite for the anti-solvent is its solving capability towards the solvent phase in which the compound of interest is dissolved. Several research groups have reported that particle size and morphology depend strongly on the nature of solvent and anti-solvent, in term of their solubility characteristics, mixing efficiency. The nature of the anti-solvent will also affect the structure of formed particles. As reported by Kakran et al. [63] changing the solvent, not only changes the supersaturation conditions, but also impact viscosity and surface tension, which also alters the nucleation rate [10, 63].

Effect of Stabilizers

The stabilization of these suspensions is of great importance for both controlling the particle size, and for developing manufacturing processes. During anti-solvent precipitation, the addition of stabilizer is known to slow the condensation and coagulation due to lower Damkhöler number (Da). Emulsifiers and surface active polymers reduce the interfacial tension and therefore, increase the nucleation rate leading to smaller particle sizes. Small molecular emulsifiers stabilize the nanodispersions usually by electrostatic or electrosteric stabilization mechanisms, while the polymers offer stabilization by steric and electrostatic repulsion. Polymer characteristics, such as rigidity, polarity, and hydrophobicity, play an important role particle growth inhibition. Furthermore, in the cases of using proteins as stabilizers, the nanoparticles have a affinity to aggregate near to their isoelectric points. In this case, other polymers or emulsifiers (e.g., pluronic F-68 or lecithin) can be used to control the precipitation step. Higher polymer or emulsifier concentrations than critical value, can induce particle aggregation via depletion flocculation mechanism [10, 70].

The combined use of emulsifiers and polymers is known to contribute to enhanced rate of particle formation, and reduce the overall particle size by the sorption of a polymer–surfactant complex on the particle surface. In multi-component stabilizer systems, the ratio of polymer either to emulsifier or active compound is also important since it influence the particle formation, stability, and loading capacity [71-74].

The results of previous studies indicated that the particle size decreased significantly with the increase of stabilizer concentration. The concentration of the stabilizer is important to the successful production of nanodispersions, which means that adequate amount of stabilizer must adsorb onto the particle surface to afford a barrier to agglomeration. The higher the concentration of the stabilizers, the further stabilizer molecules adsorbed. However, the adsorption will reach equilibrium due to the limited surface area and the effects will be reversed [75]. It is beneficial to use biopolymers to fabricate nanodispersions especially for food, agrochemical, or pharmaceutical applications. The selected biopolymers must provide a suitable matrix for preserving the properties as well as the activity of the entrapped compound, and are easy to process through the anti-solvent precipitation technique. Proteins and polysaccharides are two most common biopolymers that can be used to stabilize nanoparticles using solvent-displacement method. They are easily digested and metabolized in the human gastrointestinal tract, ensuring the release of bioactive components after ingestion. Furthermore, proteins often display anti-oxidant properties, which may be useful to protect active compounds from chemically degradation [10].

A Short Review on Literatures

Various incorporated functional lipid compounds' nanodispersions have been prepared using introduced solvent-displacement technique. The functional lipid bioactive compounds such as carotenoids, lipid soluble vitamins, sterols, water insoluble pharmaceutical drugs and lipophilic polymers are good examples in this regard. Some of these researches and their results are reviewed below.

β-carotene nanodispersions were recently prepared through solvent-displacement technique by Ribeiro et al. [76]. In this work, using environmental friendly polymers, namely, poly d,l-lactic acid (PLA) and poly d,l-lactic-co-glycolic acid (PLGA), with combination of usual small molecular emulsifier (Tween 20) or biopolymers (Gelatin), was notable. β-Carotene nanodispersions were produced by interfacial deposition of polymer, during the displacement of acetone from the dispersed phase. Their results indicated that in this kind of formulation, polymers had very high steric protective activity against the physical and chemical destabilization factors for prepared nanodispersions. Therefore, the capability of solvent-displacement technique in preparation of the most stable carotenoid nanodispersions with small particle size (80 nm), narrow particle size distribution and high encapsulating efficiency was improved in this work.

Yin et al. [74] also studied the performance of selected emulsifiers and their combinations in the preparation of β-carotene nanodispersions. In this research, the sodium caseinate (SC), Tween 20, decaglycerol monolaurate (ML750) and sucrose fatty acid ester (L1695) were selected as emulsifier and stabilizer molecules and acetone was used as solvent and organic phase. The β-carotene nanodispersions were gained in the ranges of 30-206nm, based on using emulsifier. While, SC-stabilized nanodispersions had the largest particles size, the Tween 20-stabilized ones had the smallest. The produced nanodispersions showed good physical stability during 8-week storage at over the time 4°C, but the decrease in β-carotene content in all nanodispersions over time, confirmed their low chemical stability. The SC-stabilized nanodispersions was more chemically stable, due to either the physical barrier of SC around the nano-sized β-carotene or the antioxidative activity of caseins. The combined effects of emulsifiers also were investigated. However, the prepared nanodispersions with SC-ML750 or SC-L1695 held greater mean particle sizes compared to ML750 or L1695-stabilized ones, the stability of β-carotene in these binary emulsifier systems was improved considerably. This work verified the importance of emulsifier in determining the physicochemical properties of β-carotene nanodispersions.

Bixin nanodispersions have been recently prepared using introducing nano-precipitation technique. Bixin is a good quencher of singlet oxygen, but has low solubility in water. A binary solvent system composed of 20% ethanol and 80% acetone was used as solvent system and poly-ε-caprolactone was used as stabilizing polymer. The mean particle size of gained nanodispersions were mean diameter (D4,3) of 195 ± 27 nm. The produced bixin nanodispersions showed higher thermal stability compared to free bixin [77].

Tan et al. [78] have also manufactured the lutein nanodispersions using solvent-displacement technique and acetone as suitable solvent and organic phase. They studied the effects of processing parameters, namely, addition method, addition rate, stirring time and stirring speed as well as the concentrations and structure of emulsifiers on the particle size and particle size distribution of produced nanodispersions. Their results showed that among the processing parameters, only the addition method and stirring time had significant effects ($p < 0.05$) on studied responses. The steric emulsifiers (Tween 20, 40, 60 and 80) individually or in combination with electrostatic emulsifiers (sodium dodecyl sulfate, SDS) could successfully produce the nanodispersions with particle sizes below 100 nm. However, the sodium caseinate stabilized nanodispersions with electrosteric stabilizing mechanism had larger particle sizes (140-180nm). They also reported a slight growth in mean particle size and size distribution of nanodispersions with increasing the emulsifier concentrations from 1% to 5%. Consequently, they produced stable nanodispersions with small particle sizes (72.88–142.85 nm) and narrow PSDs (polydispersity index <0.40) especially at the lowest emulsifier concentration of 0.1% with all selected emulsifiers.

Anarjan et al. [72] produced the astaxanthin nanodispersions and evaluated the effects of various processing and formulation parameters on physicochemical characteristics of gained nanodispersions. They developed a three component stabilizer system composed of small molecular emulsifier (Tween 20), protein based biopolymer (sodium caseinate) and polysaccharide based polymer (gum Arabic) in order to achieve the most desirable carotenoid nanodispersions with smallest particle size, polydispersity index and highest carotenoid loading and cellular uptake. They also studied the effects of solvent nature by selecting the dichloromethane and acetone as water immiscible and totally water miscible solvents, respectively and combinations of them to achieve various solvent polarities, on revealed response variables using a simplex centroid mixture design. They compared the ability of solvent displacement, solvent-diffusion and solvent evaporation techniques to preparation of more preferred nanodispersions. Concerning the stated goals, the astaxanthin nanodispersion prepared using a mixture of 38% dichloromethane and 62% acetone as the organic phase was shown to produce the optimum nanodispersions [50].

After that, at 2014, de Paz, et al. made a comparison between the solvent-displacement and two other common techniques, namely, high-temperature, high-pressure emulsification produced β-carotene nanodispersions. In this work, β-Carotene nanodispersions with wide range of particle sizes (120nm–135 μm) were prepared using four different originated modified n-octenyl succinate (OSA) starches as carrier materials and different procedures. Ethyl acetate and ethanol were the selected solvents which supports the mentioned preparation techniques. The smaller particle size of ethanol produced nanodispersions compared to ethyl acetate ones, confirmed the proficiency of solvent-displacement compared to emulsification techniques in production of more desired organic nanodispersions. However, ethanol yielded the nanodispersions with less loading of lipoid active compounds, maybe due to less solubility of β-carotene in ethanol compared to ethyl acetate. They also showed that increasing the β-carotene concentration in system caused an increase on particle sizes of produced nanodispersions. Two process parameters, namely, concentration of emulsifier and organic solvent–water ratio were also studied in this research. The results indicated that an increase in the concentration of emulsifier produced an increase in the encapsulation efficiency, and at high organic solvent–water ratios, a drastic increase in the particle size was observed, independently of the OSA-starch used. The produced nanoparticles were stable after four months of storage [79].

α-Tocopherol nanodispersions were produced recently, using acetone as organic phase and the mixing parameters were optimized for producing the most desirable product with the smallest particle size, size distribution and highest α-tocopherol retention of nanodispersions. Second order regression models, with high coefficient of determination values (R 2 > 0.94), were proposed for predicting the variation of selected response parameters with mixing factors. The optimum mixing speed and time were reported as 380 rpm and 70 second, respectively, in this research [80].

Furthermore, Campardelli and Roverchon [81] have also developed a supercritical assisted solvent-displacement process, named SAILA (Supercritical Assisted Injection in Liquid Anti-solvent) for the production of α-tocopherol nanodispersions. In this process, an expanded liquid solution was formed by supercritical CO2, an organic solvent (acetone, ethanol, isopropanol) and solid solute (α-tocopherol). Then, the solution was depressurized directly through an injector nuzzle into water containing suitable emulsifier (Tween 80). Therefore, the water based solution worked as a liquid anti-solvent. They also evaluated the effects of process parameters such as kind of solvent–anti-solvent, concentrations of solute and emulsifier, expanded liquid temperature, pressure and composition on particle size, size distribution and zeta potential. Thy found that using acetone as solvent, with smaller injecting nuzzle diameter and less expanded liquid pressure and higher expansion temperature led to production of the smallest nanodispersions (220 nm). Increasing the concentration of α-tocopherol caused an increase in particle size and size distribution of system. Furthermore, an optimum concentration for emulsifier was observed in order to gain the smallest nanodispersions. Therefore, SAILA process was successfully employed for the production of α-tocopherol nanodispersions. The produced water dispersed nanoparticles were spherical and stable during storage at 4°C for 30 days.

The similar process was also employed for preparation of poly-ethylene glycol (PEG) stabilized β-carotene and lutein nanodispersions [82]. Therefore, the co-precipitation of PEG and carotenoids was presented. CO2 in the form of supercritical was used as solvent in this research and the effects of main process and formulation parameters was evaluated subsequently. The lutein nanodispersions owned smaller particle size with better coverage of PEG compared to β-carotene nanodispersions. The polymer/carotenoid ratio was found as the most effective parameter on the morphology characteristics of produced carotenoid nanodispersions.

The polymer nanodispersions can also be prepared using introduced technique. For example, Birnbaum et al. [83] have studied the production of water dispersed nanoparticles of poly(lactic acid-co-glycolic acid) (PLAGA), which are excellent compounds for drug delivery uses, using several combinations of organic solvents and emulsifiers. The mean particle sizes of produced PLAGA nanodispersions were ranged from 65 to 466nm. Their results showed that using either acetone or ethyl acetate as the organic solvent and poly(vinyl alcohol) or human serum albumin as the emulsifier, can produce the most desirable polymeric nanodispersions.

Poly-ε-caprolactone (PCL) nanodispersions were also produced in recent decades. Two different PCL molecular weights have been employed, using acetone as solvent and and water as anti-solvent. The thermodynamic and kinetic parameters, such as solubility and interfacial tension of PCL in water-acetone, were determined and the effects of the process parameters on mean particle size, zeta potential and size distribution of produced nanodispersions were evaluated. Results clearly indicated the strong effects of mixing parameters on the particle size and size distribution of PCL-nanodispersions [84].

Polysulfone (PSF) nanodispersions were also prepared through rapid solvent displacement in a membrane contactor between solvent (dimethylformamide, DMF) and non-solvent (water) phases. In this nano-precipitation process, despite of instant nucleation process, their growth and aggregation were considerably slow. Their results showed an enhancement in particle sizes of produced polymer nanodispersions due to reservation of DMF in system. Therefore, an improved nano-precipitation method appending a quenching stage was offered. It was also resulted that introducing the ethyl acetate could lead to successfully production of well-dispersed PSF nanoparticles. The operating factors, including the PSF concentration in the dispersed fluid and the ratio of two phases had significant effects on the characteristics of gained product [85].

The water dispersed Poly(lactide-co-glycolide) (PLGA) nanoparticles have also been prepared recently for drug delivery applications. The effects of organic phase (solvent nature) on status of "Ouzo region" were evaluated as well. The used solvents were tetrahydrofuran (THF), 1,4-dioxane, acetone and dimethyl sulfoxide (DMSO). The results showed a considerable shift for "Ouzo region" by increasing the polymer concentration and changing the solvent (rank order: THF < 1,4-dioxane < acetone < DMSO). The DMSO and THF produced the smallest (300 nm) and the largest (800 nm) nanoparticles. By manipulating the polymer concentration, production of smaller nanoparticles (40-200 nm) was found to be possible using selected solvents. It was concluded that "Ouzo diagrams" can be a valuable tool for size reduction and drug delivery researches to find the operating area for the production of stable colloidal formulations by the solvent displacement or nano-precipitation technique [86].

Lipophilic prescriptions are another group of compounds that can be converted to nano-sized scales via described process. Cyclosporin A (CyA) is a good sample in this regard. It is suitable candidate for incorporation in nanodispersion systems in order to reduce its adverse effects due to its use under conventional pharmaceutical dosage forms and increase bioavailability after oral administration. The CyA nanodispersions were already produced by Molpeceres et al. [87]. Temperature of the aqueous phase, volume of the organic phase, injection needle diameter and the used polymer and emulsifier concentrations were selected as the most important independent variables in their research. They found that the polymer and emulsifier concentrations, and the diameter of the needle used, significantly affected the particle size, size distribution and drug encapsulation efficiency.

Lopalco et al. [88] have also encapsulated the Oxcarbazepine in nanodispersion systems to improve its brain delivery and decrease its adverse effects. The poly(lactide-co-glycolide) (PLGA) with or without emulsifier was used as biocompatible stabilizing molecules. The produced nanodispersions was amorphous with 69% encapsulation efficiency, 140-170 nm mean particle sizes, below 0.3 PDI, and less than −34 mV zeta potential. Transport studies by fluorescently-labeled nanoparticles confirmed higher permeability of emulsifier-coated nanoparticles compared to others.

CONCLUSION

Solvent-displacement or anti-solvent precipitation technique was reviewed in this chapter as valuable, cost effective, efficient, scalable technique for producing water dispersed functional lipid bioactive compounds' nanoparticles (nanodispersions). This bottom-up process is able to produce nanodispersions for several food, agriculture and pharmaceutical applications, to improve their safety and functionality by entrapping, protecting and controlling their release. The nanoparticle formation occurred in two steps, namely, nucleation and particle growth with distinct mechanisms. The processing and formulation parameters can affect the characteristics of produced nanodispersions by adjusting their size, shape, structure, surface charge, and etc. Due to increasing application of produced nanodispersions, their effects either on human cells or ecological systems are becoming more important. They can be potentially unsafe, because of their ability to cross cellular layers or accumulate in some tissues. Therefore, long-term effects of exposure to nanoparticles should be carefully evaluated.

REFERENCES

[1] Rowe JM, Johnston KP. Precipitation technologies for nanoparticle production. Formulating Poorly Water Soluble Drugs: Springer; 2012. p. 501-68.

[2] Nkansah P, Antipas A, Lu Y, Varma M, Rotter C, Rago B, et al. Development and evaluation of novel solid nanodispersion system for oral delivery of poorly water-soluble drugs. *Journal of Controlled Release.* 2013;169:150-61.

[3] Lepeltier E, Bourgaux C, Couvreur P. Nanoprecipitation and the "Ouzo effect": Application to rug delivery devices. *Advanced Drug Delivery Reviews.* 2014;71:86-97.

[4] Fessi H, Puisieux F, Devissaguet JP, Ammoury N, Benita S. Nanocapsule formation by interfacial polymer deposition following solvent displacement. *International Journal of Pharmaceutics.* 1989;55:R1-R4.

[5] Bilati U, Allémann E, Doelker E. Development of a nanoprecipitation method intended for the entrapment of hydrophilic drugs into nanoparticles. *European Journal of Pharmaceutical Sciences.* 2005;24:67-75.

[6] Ahlin Grabnar P, Kristl J. The manufacturing techniques of drug-loaded polymeric nanoparticles from preformed polymers. *Journal of Microencapsulation.* 2011;28:323-35.

[7] Freitag K. Preparation of nanodispersions by antisolvent precipitation: lmu; 2010.

[8] Panigrahi S. Studies on Growth Kinetics of Nanoparticles Formation In Bulk Solution: National Institute of Technology, Rourkela; 2010.

[9] Thanh NT, Maclean N, Mahiddine S. Mechanisms of nucleation and growth of nanoparticles in solution. *Chemical Reviews.* 2014;114:7610-30.

[10] Joye IJ, McClements DJ. Production of nanoparticles by anti-solvent precipitation for use in food systems. *Trends in Food Science and Technology.* 2013;34:109-23.

[11] Destrée C, Nagy J. Mechanism of formation of inorganic and organic nanoparticles from microemulsions. *Advances in Colloid and Interface Science.* 2006;123:353-67.

[12] Prathna T, Chandrasekaran N, Raichur AM, Mukherjee A. Kinetic evolution studies of silver nanoparticles in a bio-based green synthesis process. *Colloids and Surfaces A: Physicochemical and Engineering Aspects.* 2011;377:212-6.

[13] D'Addio SM, Prud'homme RK. Controlling drug nanoparticle formation by rapid precipitation. *Advanced Drug Delivery Reviews.* 2011;63:417-26.

[14] Lindenberg C. Optimizing the precipitation of organic compounds: Diss., Eidgenössische Technische Hochschule ETH Zürich, Nr. 18329, 2009; 2009.

[15] Matteucci ME, Hotze MA, Johnston KP, Williams RO. Drug nanoparticles by antisolvent precipitation: mixing energy versus surfactant stabilization. *Langmuir.* 2006;22:8951-9.

[16] Beck C, Dalvi SV, Dave RN. Controlled liquid antisolvent precipitation using a rapid mixing device. *Chemical Engineering Science.* 2010;65: 5669-75.

[17] Desai C, Meng X, Yang D, Wang X, Akkunuru V, Mitra S. Effect of solvents on stabilization of micro drug particles. *Journal of Crystal Growth.* 2011;314:353-8.

[18] Horn D, Rieger J. Organic nanoparticles in the aqueous phase—theory, experiment, and use. *Angewandte Chemie International Edition.* 2001;40:4330-61.

[19] Sinha B, Müller RH, Möschwitzer JP. Bottom-up approaches for preparing drug nanocrystals: formulations and factors affecting particle size. International *Journal of Pharmaceutics.* 2013;453:126-41.

[20] Deshmukh K, Ahmad J, Joshi G, Ahamed MB, Hägg MB. Stability and electrokinetic properties of aqueous TiO2 nanoparticles dispersion in polyallylamine and polyvinyl alcohol blend systems. *Journal of Polymer Research.* 2014;21:1-9.

[21] Labille J, Brant J. Stability of nanoparticles in water. *Nanomedicine.* 2010;5:985-98.

[22] Elimelech M, Gregory J, Jia X, Williams R. Particle Deposition and Aggregation: Measurement, Modelling and Simulation 1995. Butterworth-Heinemann: Oxford, UK.

[23] Peltonen L, Hirvonen J. Pharmaceutical nanocrystals by nanomilling: critical process parameters, particle fracturing and stabilization methods. *Journal of Pharmacy and Pharmacology.* 2010;62:1569-79.

[24] Mullaugh KM, Luther III GW. Growth kinetics and long-term stability of CdS nanoparticles in aqueous solution under ambient conditions. *Journal of Nanoparticle Research.* 2011;13:393-404.

[25] Israelachvili JN. Intermolecular and surface forces: revised third edition: Academic press; 2011.

[26] Rouxel D, Hadji R, Vincent B, Fort Y. Effect of ultrasonication and dispersion stability on the cluster size of alumina nanoscale particles in aqueous solutions. *Ultrasonics Sonochemistry.* 2011;18:382-8.

[27] Elgart A, Cherniakov I, Aldouby Y, Domb AJ, Hoffman A. Lipospheres and pro-nano lipospheres for delivery of poorly water soluble compounds. *Chemistry and Physics of Lipids.* 2012;165:438-53.

[28] Cheng JC, Fox RO. Kinetic modeling of nanoprecipitation using CFD coupled with a population balance. *Industrial and Engineering Chemistry Research.* 2010;49:10651-62.

[29] Pirkle C, Foguth LC, Brenek SJ, Girard K, Braatz RD. Computational fluid dynamics modeling of mixing effects for crystallization in coaxial nozzles. Chemical Engineering and Processing: Process Intensification. 2015.

[30] Choi Y-J, Chung S-T, Oh M, Kim H-S. Investigation of crystallization in a jet Y-mixer by a hybrid computational fluid dynamics and process simulation approach. *Crystal Growth and Design.* 2005;5:959-68.

[31] Woo XY, Tan RB, Braatz RD. Modeling and computational fluid dynamics− population balance equation− micromixing simulation of impinging jet crystallizers. *Crystal Growth and Design.* 2008;9:156-64.

[32] Woo XY, Tan RB, Chow PS, Braatz RD. Simulation of mixing effects in antisolvent crystallization using a coupled CFD-PDF-PBE approach. *Crystal Growth and Design.* 2006;6:1291-303.

[33] Mahajan AJ, Kirwan DJ. Nucleation and growth kinetics of biochemicals measured at high supersaturations. *Journal of Crystal Growth.* 1994;144:281-90.

[34] Stainmesse S, Fessi H, Devissaguet J-P, Puisieux F, Theis C. Process for the preparation of dispersible colloidal systems of a substance in the form of nanoparticles. Google Patents; 1992.

[35] Anandharamakrishnan C. Liquid-Based Nanoencapsulation Techniques. Techniques for Nanoencapsulation of Food Ingredients: Springer; 2014. p. 29-41.

[36] Quintanar-Guerrero D, Allémann E, Fessi H, Doelker E. Preparation techniques and mechanisms of formation of biodegradable nanoparticles from preformed polymers. *Drug Development and Industrial Pharmacy.* 1998;24:1113-28.

[37] Koch CC. Nanostructured materials: processing, properties and applications: William Andrew; 2006.

[38] Gassmann P, List M, Schweitzer A, Sucker H. Hydrosols: alternatives for the parenteral application of poorly water soluble drugs. *European Journal of Pharmaceutics and Biopharmaceutics.* 1994;40:64-72.

[39] Lindfors L, Skantze P, Skantze U, Rasmusson M, Zackrisson A, Olsson U. Amorphous drug nanosuspensions. 1. Inhibition of Ostwald ripening. *Langmuir.* 2006;22:906-10.

[40] Thorat AA, Dalvi SV. Liquid antisolvent precipitation and stabilization of nanoparticles of poorly water soluble drugs in aqueous suspensions: Recent developments and future perspective. *Chemical Engineering Journal.* 2012;181:1-34.

[41] Keck CM, Müller RH. Drug nanocrystals of poorly soluble drugs produced by high pressure homogenisation. *European Journal of Pharmaceutics and Biopharmaceutics*. 2006;62:3-16.

[42] Rogers TL, Gillespie IB, Hitt JE, Fransen KL, Crowl CA, Tucker CJ, et al. Development and characterization of a scalable controlled precipitation process to enhance the dissolution of poorly water-soluble drugs. *Pharmaceutical Research*. 2004;21:2048-57.

[43] Chan H-K, Kwok PCL. Production methods for nanodrug particles using the bottom-up approach. *Advanced Drug Delivery Reviews*. 2011; 63:406-16.

[44] Hatkar UN, Gogate PR. Process intensification of anti-solvent crystallization of salicylic acid using ultrasonic irradiations. *Chemical Engineering and Processing: Process Intensification*. 2012;57:16-24.

[45] Lonare AA, Patel SR. Antisolvent crystallization of poorly water soluble drugs. *International Journal of Chemical Engineering Applications*. 2013;4:337-41.

[46] Pal SL, Jana U, Manna P, Mohanta G, Manavalan R. Nanoparticle: an overview of preparation and characterization. *Journal of Applied Pharmaceutical Science*. 2011;1:228-34.

[47] Ubrich N, Bouillot P, Pellerin C, Hoffman M, Maincent P. Preparation and characterization of propranolol hydrochloride nanoparticles: a comparative study. *Journal of Controlled Release*. 2004;97:291-300.

[48] Anarjan N, Mirhosseini H, Baharin BS, Tan CP. Effect of processing conditions on physicochemical properties of astaxanthin nanodispersions. *Food Chemistry*. 2010;123:477-83.

[49] Anarjan N, Nehdi IA, Tan CP. Influence of astaxanthin, emulsifier and organic phase concentration on physicochemical properties of astaxanthin nanodispersions. *Chemistry Central Journal*. 2013;7:127.

[50] Anarjan N, Tan CP, Ling TC, Lye KL, Malmiri HJ, Nehdi IA, et al. Effect of organic-phase solvents on physicochemical properties and cellular uptake of astaxanthin nanodispersions. *Journal of Agricultural and Food Chemistry*. 2011;59:8733-41.

[51] Kikic I, Lora M, Bertucco A. A thermodynamic analysis of three-phase equilibria in binary and ternary systems for applications in rapid expansion of a supercritical solution (RESS), particles from gas-saturated solutions (PGSS), and supercritical antisolvent (SAS). *Industrial and Engineering Chemistry Research*. 1997;36:5507-15.

[52] Djerafi R, Masmoudi Y, Crampon C, Meniai A, Badens E. Supercritical anti-solvent precipitation of ethyl cellulose. *The Journal of Supercritical Fluids*. 2015.

[53] Ezhilarasi P, Karthik P, Chhanwal N, Anandharamakrishnan C. Nanoencapsulation techniques for food bioactive components: a review. *Food and Bioprocess Technology*. 2013;6:628-47.

[54] Sekhon BS. Supercritical fluid technology: an overview of pharmaceutical applications. *International Journal of Pharm Tech Research*. 2010;2:810-26.

[55] Nagavarma B, Yadav HK, Ayaz A, Vasudha L, Shivakumar H. Different techniques for preparation of polymeric nanoparticles—a review. *Asian Journal of Pharmaceutical and Clinic Research*. 2012;5:16-23.

[56] Meziani MJ, Pathak P, Hurezeanu R, Thies MC, Enick RM, Sun YP. Supercritical-Fluid Processing Technique for Nanoscale Polymer Particles. *Angewandte Chemie International Edition*. 2004;43:704-7.

[57] Thakur R, Gupta RB. Rapid expansion of supercritical solution with solid cosolvent (RESS-SC) process: formation of griseofulvin nanoparticles. *Industrial and Engineering Chemistry Research.* 2005;44:7380-7.

[58] Martín A, Cocero MJ. Micronization processes with supercritical fluids: fundamentals and mechanisms. *Advanced Drug Delivery Reviews.* 2008;60:339-50.

[59] Prud'homme R, Saad W, Mayer L. Paclitaxel conjugate block copolymer nanoparticle formation by flash nanoprecipitation. *NSTI Nanotech.* 2006:824-6.

[60] Albano AA, Phuapradit W, Sandhu HK, Shah NH. Complexes significantly increase the bioavailability of poorly soluble therapeutically active compounds. Google Patents; 2002.

[61] Flaherty KT, Puzanov I, Kim KB, Ribas A, McArthur GA, Sosman JA, et al. Inhibition of mutated, activated BRAF in metastatic melanoma. *New England Journal of Medicine.* 2010;363:809-19.

[62] Kitamura M, Sugimoto M. Anti-solvent crystallization and transformation of thiazole-derivative polymorphs—I: Effect of addition rate and initial concentrations. *Journal of Crystal Growth.* 2003;257:177-84.

[63] Kakran M, Sahoo N, Tan IL, Li L. Preparation of nanoparticles of poorly water-soluble antioxidant curcumin by antisolvent precipitation methods. *Journal of Nanoparticle Research.* 2012;14:1-11.

[64] Langer K, Balthasar S, Vogel V, Dinauer N, von Briesen H, Schubert D. Optimization of the preparation process for human serum albumin (HSA) nanoparticles. *International Journal of Pharmaceutics.* 2003;257: 169-80.

[65] O'Grady D, Barrett M, Casey E, Glennon B. The effect of mixing on the metastable zone width and nucleation kinetics in the anti-solvent crystallization of benzoic acid. *Chemical Engineering Research and Design.* 2007;85: 945-52.

[66] Yu Z, Tan R, Chow P. Effects of operating conditions on agglomeration and habit of paracetamol crystals in anti-solvent crystallization. *Journal of Crystal Growth.* 2005;279:477-88.

[67] Guo Z, Zhang M, Li H, Wang J, Kougoulos E. Effect of ultrasound on anti-solvent crystallization process. *Journal of Crystal Growth.* 2005;273:555 63.

[68] Sheng F, Chow PS, Dong Y, Tan RB. Preparation of β-carotene nanoparticles by antisolvent precipitation under power ultrasound. *Journal of Nanoparticle Research.* 2014;16:1-9.

[69] De Castro ML, Priego-Capote F. Ultrasound-assisted crystallization (sonocrystallization). *Ultrasonics Sonochemistry.* 2007;14:717-24.

[70] Dalvi SV, Dave RN. Controlling particle size of a poorly water-soluble drug using ultrasound and stabilizers in antisolvent precipitation. *Industrial and Engineering Chemistry Research.* 2009;48:7581-93.

[71] Meng X, Chen Y, Chowdhury SR, Yang D, Mitra S. Stabilizing dispersions of hydrophobic drug molecules using cellulose ethers during anti-solvent synthesis of micro-particulates. *Colloids and Surfaces B: Biointerfaces.* 2009;70:7-14.

[72] Anarjan N, Tan CP. Developing a three component stabilizer system for producing astaxanthin nanodispersions. *Food Hydrocolloids.* 2013;30:437-47.

[73] Anarjan N, Tan CP. Chemical stability of astaxanthin nanodispersions in orange juice and skimmed milk as model food systems. *Food Chemistry.* 2013;139:527-31.

[74] Yin L-J, Chu B-S, Kobayashi I, Nakajima M. Performance of selected emulsifiers and their combinations in the preparation of β-carotene nanodispersions. *Food Hydrocolloids*. 2009;23:1617-22.

[75] Xu Y, Liu X, Lian R, Zheng S, Yin Z, Lu Y, et al. Enhanced dissolution and oral bioavailability of aripiprazole nanosuspensions prepared by nanoprecipitation/ homogenization based on acid–base neutralization. *International Journal of Pharmaceutics*. 2012;438:287-95.

[76] Ribeiro HS, Chu B-S, Ichikawa S, Nakajima M. Preparation of nanodispersions containing β-carotene by solvent displacement method. *Food Hydrocolloids*. 2008;22:12-7.

[77] Lobato KBdS, Paese K, Forgearini JC, Guterres SS, Jablonski A, Rios AdO. Evaluation of stability of bixin in nanocapsules in model systems of photosensitization and heating. *LWT - Food Science and Technology*. 2015;60:8-14.

[78] Tan TB, Yussof NS, Abas F, Mirhosseini H, Nehdi IA, Tan CP. Forming a lutein nanodispersion via solvent displacement method: The effects of processing parameters and emulsifiers with different stabilizing mechanisms. *Food Chemistry*. 2016;194:416-23.

[79] de Paz E, Martín Á, Bartolomé A, Largo M, Cocero MJ. Development of water-soluble β-carotene formulations by high-temperature, high-pressure emulsification and antisolvent precipitation. *Food Hydrocolloids*. 2014;37:14-24.

[80] Anarjan N, Jaberi N, Yeganeh-Zare S, Banafshehchin E, Rahimirad A, Jafarizadeh-Malmiri H. Optimization of Mixing Parameters for α-Tocopherol Nanodispersions Prepared Using Solvent Displacement Method. *Journal of the American Oil Chemists Society*. 2014;91:1397-405.

[81] Campardelli R, Reverchon E. α-Tocopherol nanosuspensions produced using a supercritical assisted process. *Journal of Food Engineering*. 2015;149:131-6.

[82] Martín A, Mattea F, Gutiérrez L, Miguel F, Cocero MJ. Co-precipitation of carotenoids and bio-polymers with the supercritical anti-solvent process. *The Journal of Supercritical Fluids*. 2007;41:138-47.

[83] Birnbaum DT, Kosmala JD, Brannon-Peppas L. Optimization of preparation techniques for poly(lactic acid-co-glycolic acid) nanoparticles. *Journal of Nanoparticle* Research. 2000;2:173-81.

[84] Lince F, Marchisio DL, Barresi AA. Strategies to control the particle size distribution of poly-ε-caprolactone nanoparticles for pharmaceutical applications. *Journal of Colloid and Interface Science*. 2008;322:505-15.

[85] Liu Y, Lu YC, Luo GS. Modified nanoprecipitation method for polysulfone nanoparticles preparation. *Soft Matter*. 2014;10:3414-20.

[86] Beck-Broichsitter M, Nicolas J, Couvreur P. Solvent selection causes remarkable shifts of the "Ouzo region" for poly(lactide-co-glycolide) nanoparticles prepared by nanoprecipitation. *Nanoscale*. 2015;7:9215-21.

[87] Molpeceres J, Guzman M, Aberturas MR, Chacon M, Berges L. Application of central composite designs to the preparation of polycaprolactone nanoparticles by solvent displacement. *Journal of Pharmaceutical Sciences*. 1996;85:206-13.

[88] Lopalco A, Ali H, Denora N, Rytting E. Oxcarbazepine-loaded polymeric nanoparticles: development and permeability studies across in vitro models of the blood–brain barrier and human placental trophoblast. *International Journal of Nanomedicine*. 2015;10 1985-1996.

In: High Value Processing Technologies
Editors: A. Berenjian, H. Jafarizadeh-Malmiri et al.
ISBN: 978-1-63484-453-6

Chapter 3

CONTINUOUS MANUFACTURING OF PHARMACEUTICAL AND BIOPHARMACEUTICAL AGENTS, PRODUCING MORE WITH LESS

Nima Yazdanpanah [*]
Massachusetts Institute of Technology, Cambridge, Massachusetts, US

ABSTRACT

Continuous manufacturing of pharmaceutical and biopharmaceutical compounds is an emerging process manner in order to enable producing drugs in a chipper, faster, more efficient, with better quality, with lower footprint, and more flexible way. Although the continuous manufacturing concept is not new and already in practice for chemical, food, petrochemicals, etc. industries, the classic and dominant batch process is almost the only current fashion in pharma and biopharma industry. Although continuous processing is more complicated to design and operate compared to batch, the significant advantages and high investment return rate make the "revolution" inevitable. More recent developments have been focused on making continuous processing principles accessible to the industry by using familiar unit operations operated in a continuous fashion. However, overall process design, and system engineering toward an end-to-end continuous process is the significant leapfrogging, which makes this enabling technology successful. Therefore, from the control and flow points of view, a continuous process will be a single unit operation which could include several batch or semi-continuous units. In this chapter advantages, challenges, implementation and regulation aspects of continuous manufacturing will be discussed.

WHY CONTINUOUS MANUFACTURING?

For both upstream and downstream steps of pharma and biopharma processing, the classic and largely predominant approach remains batch processing. The continuous

[*] Corresponding author: Email: nima@mit.edu.

manufacturing is not a new concept, and has been practiced in chemical, petrochemical, food, pulp and paper industry for many years. But pharmaceutical and biopharmaceutical compounds are still vastly being produced in batch. Recently continuous manufacturing is being adopted or considered for many new drug-manufacturing processes, and higher future adoption is being expected for new production lines, and upgrade of the “old” fashion batch processes. Leading adopters, including Novartis, Pfizer, Genzym, Amgen, Byer, and Biogen are working on developing new continuous manufacturing for their upstream and downstream processes, and very few end-to-end fully continuous processes are going to be implemented in the industry [1]. Most of the current proposed configurations are hybrid systems of continuous upstream and batch downstream or vice versa. With recent overcoming of challenges and barriers (including regulatory concerns) and development of advance PAT, monitoring and control tools, the trend is toward the fully continuous end-to-end platforms and integration of entire chemical unit operation chains.

Continuous processing is more complicated to design and operate compared to batch; process scalability, reliability, and consistency can be problematic and process validation and regulatory acceptance are not straightforward. Recent advancements in flow-chemistry (for drug synthesis and upstream of pharma), and upstream bioprocessing (such as perfusion bioreactors), have had significant impact toward fulfillment of the mission. The industries have strong technical and economical motivations in pursue of continuous manufacturing, such as maintaining consistent high quality with robust design and control; cost effectiveness by attaining high productivity and lower capital investment, and lower operation cost, down-time, and foot-print; standardization with a manufacturing model that can be used for multi compound manufacturing and for multiple facilities. In general continuous process technologies can increase productivity, resolve facility utilization and facility fit challenges, reduce capital expenditures, improve product quality, and ensure manufacturing consistency.

The batch process has been adopted in the pharmaceutical industry mainly for historical reasons; it is easier to define a batch, it is easier to capture all the events in the batch history, and it is easier to operate and control [2, 3]. Batch processes require frequent downtime between the batches to turnaround the equipment and they are not as efficient as a continuous process in terms of process yields [4]. A batch process is run for a short time and a lot of time is spent between the batches for termination of the previous batch, cleaning, and reconditioning of the equipment (turnaround), and preparation and build-up of the process intermediates for the next batch. The time period between the batches, referred to as downtime, is not used to manufacture the product and it impacts the cycle time and plant utilization efficiency. An internal study in MIT-Novartis Center for Continuous Manufacturing shows that the continuous manufacturing could bring: reduction of asset footprint (40-90%), reduction in capital expenditure (25-60%), reduction of operational costs (25-60%), and reduction of raw material and intermediate inventories. These advantages are highly lucrative for any batch processing plant to invest in this kind of research and development toward continuous manufacturing. These benefits being achieved through:

1) Increased productivity: The number of operator required will be decreased. Continuous reactors eliminate the need for buffer tanks and transporting large amount of mid products. In bioprocessing much higher bioreactor cell densities can be attained. Continuous processes run at their maximum capacity and at a steady

state while a batch process goes through a cycle where only a portion of the run is fully effective. These provide higher product yield and concentration.

2) Reduced costs: Continuous operation allows use of much smaller-scale equipment. With a smaller volume reactor and bioreactor, operating over time provides as much product as larger batch and fed-batch operating equipment. Smaller equipment cost less and requires less utilities, space, facilities, and foot-print.
3) Increased flexibility: Continuous manufacturing enables more efficient facility utilization. Such operations can be modulated, intensified, portable and clonable.
4) Improved quality: On the steady-state condition of the processes, the flocculation of concentration (feed and product) is eliminated. The batch-to-batch quality variation is minimized, and if any problem occurs, only limited part of the product, not the entire, can be rejected.

Continuous manufacturing had its modern roots in the idea of process intensification, which means more material transformation in less reaction space, which in turn leads to the big technical benefits: better process control through stringent enforcement of process conditions on a microscopic scale and ultimately the option for smaller equipment. In order to get the most intense processes, the need is to minimize the reaction space thus maximize the concentration of the transformations over space. In order to eliminate loading and unloading times of the reactor that do not add to the transformation as such and hence do not contribute value, a simultaneous inlet and outlet of materials is needed. As a consequence of this approach, all elements as there are inlet of material, transformation and outlet of material are then theoretically without discrete elements, except for start and finish, hence the process is truly continuous and can be operated at any desired length of time.

Together with the fact that continuous manufacturing is the ultimate in lean manufacturing, flexibility is a key aspect, either in meeting a need to produce in the same line product with different content of active pharmaceutical ingredients (API) or different products or different volumes during the year. This is an important evaluation element for the design of a continuous line, in terms of size (capacity) and modularity (different products) with the possibility to adopt different paths for the product through various modules according to the formulation chosen. Consequently if it is needed to change production on a line, the cleanibility and the setup time become extremely important, cleanup including waste at the beginning and end of production. The need for flexibility is likely to influence the level of continuity of the system, in terms of integration between upstream and downstream.

ECONOMIC BENEFIT OF CONTINUOUS MANUFACTURING

To date, the companies' view is that there have been two different main drivers for investment in continuous manufacturing:

- Creating suitable platforms for future drug development with an expectation of future benefits.
- Investment based on a current business case with "near term" payback.

Table 1. Example: Summary of CapEx Differences for continuous manufacturing, Relative to Batch Case [6]

Cost of KI:	API mass loading = 10%			API mass loading = 50%		
	$100/kg	**$500/kg**	**$3000/kg**	**$100/kg**	**$500/kg**	**$3000/kg**
Batch (basis for differences)	[$315M]	[$346M]	[$542M]	[$429M]	[$585M]	[$1565M]
CM with direct tablet formation	–31%	–36%	–55%	–42%	–55%	–76%
CM with roller compaction	–23%	–29%	–50%	–36%	–50%	–74%

Hence, the recent traction in secondary processing has been driven by the creation of smaller scale equipment, which has been specifically designed in order to minimize the amount of material required during development.

The cost of lab-scale equipment may be considered high. It includes, however, all of the experience, training, and continuous support provided by the supplier, which is much higher in the case of emerging technology than for a conventional piece of equipment [5]. A review of the fine and commodity chemical industries demonstrates that continuous manufacturing could offer both operating expenditure (OpEx) and capital expenditure (CapEx) savings for the pharmaceutical industry. The most significant process economic impacts anticipated for those who adopt continuous processing for pharmaceuticals and biopharmaceuticals are based on process intensification, that is, the ability to produce more product with less capital and in a smaller footprint. This ability will result in lower capital costs for construction of the facility and purchase of equipment, lower energy and water usage, and fewer overheads related to facility footprint. There are additional related benefits, including the ability to use single-use equipment to produce significantly larger quantities of material than could be achieved in single-use batch mode.

Table 1 shows a study of CapEx differences for two of continuous manufacturing technique versus a batch model. Table 2 show a result of a study for Annual OpEx Differences for continuous manufacturing, Relative to Batch Case. The results show significant reduction in the CapEx and OpEx for different scenarios, which would be highly lucrative for industry to invest in continuous manufacturing.

Table 2. Example: Summary of Annual OpEx Differences for continuous manufacturing, Relative to Batch Case [6]

Cost of KI:	API mass loading = 10%			API mass loading = 50%		
	$100/kg	**$500/kg**	**$3000/kg**	**$100/kg**	**$500/kg**	**$3000/kg**
batch (basis for differences)	[$136M]	[$226M]	[$785M]	[$531M]	[$979M]	[$3777M]
CM1 with direct tablet formation	–19%	–6%	8%	–22%	–6%	9%
CM1 with roller compaction	–19%	–6%	8%	–22%	–6%	9%

Table 3. Example: Comparison of a batch and continuous downstream process to purify 30 kg of monoclonal antibody [9]

	Batch	Continuous (SMB)
Resin Volume (l)	460	100
Column size	140x30 cm	40X50 cm
Cycles/Column	3 cycles	16 columns
Processes time (h)	21	20
Productivity (g/l.h)	2.9	10.4

In addition to these generalized benefits for reducing operating costs, certain unit operations have specific economic benefits that will be derived from continuous operation. For example, operating chromatography in a multicolumn semi-continuous mode, using simulated moving bed technology or periodic countercurrent chromatography will result in lower media and buffer requirements due to the higher column loadings that can be achieved with equivalent yields using these modes of operation [7, 8]. Table 3 compares a batch and continuous downstream process to purify 30 kg of monoclonal antibody. The performance shows five times higher productivity for one fifth of resin volume for continuous manufacturing versus the batch mode.

EFFECTS ON SUPPLY CHAIN

Future supply chains will be required to enhance affordability and availability for patients and healthcare providers alike despite the increased product complexity. The potential opportunities for 'continuous processing' centric manufacturing supply chains are significant. It provides new capabilities for firms to meet as yet unmet patient needs, by developing capabilities to supply niche markets that are currently uneconomical to serve due to small product volumes linked to specific patient populations. The volume flexibility afforded by continuous processing, unconstrained by batch size, has major implications for materials requirements and inventory, shortening dramatically the shelf-life of products that are often determined by minimum batch volumes.

Better adaption to market supply and country versions is a direct benefit of the technology. A variety of small markets (in terms of units produced, not value) can drive a substantial accumulated volume. Impact of individualized medicine can require a variety of strengths that need to be produced, each as multiple country versions. In the batch world of today this is difficult to handle and negatively impacts on the Cost of Goods, inventory and shelf life requirements, as well as operational costs linked to increased complexity [10].

Late customization and deferment models, is another alternative supply scenario, where final product definition is achieved in-market within a distributed and geographically dispersed 'final stage' manufacturing activity; in this scenario both upstream and downstream stages may preferentially support continuous process technologies that support 'material flow' supply dynamics as opposed to the current 'batch' campaign model. The opportunities for a significant reduction in inventory through continuous processing results from chemical processing models offering reduced process steps and process equipment that provides significant volume flexibility. This leads to substantial reductions in inventory which in-turn enable moving to more of a 'demand driven' replenishment model rather than the current long-term forecasting approaches with wider opportunities for manufacturing and supply chain integration. It also enable the ability to operate on substantially shorter lead times for product replenishment by significantly reducing intermediate and finished goods stock levels. In addition, consideration of chemical routes involving short-lived unstable intermediates within continuous manufacturing, normally avoided in batch-chemistry, opening up alternative synthesis routes.

To analyze the impact of continuous manufacturing on the supply chain, there would be multiple supply chain models with configuration scenarios that could be considered. One of the core elements is the geographically dispersed production networks, supported by more repeatable continuous based production processes, and offering significant volume flexibility with a flow-through demand driven supply dynamic, progressively replacing the 'batch campaign.' Unit operations will involve fewer production steps that change production dynamics from multi-stage multi-location to single location processing

Multiple supply chain models that support different levels of geographical reach, with e.g., centralized supply solutions feeding late customized/ consolidation models; dispersed supply models that support local near-market replenishment models. Small continuous lines enable localization in small markets, where the Factory in a box is one scenario but so are smaller more standard factory operations that are substantially less capital, labor and energy intensive. Manufacturing 'on-demand' with less inventory enabled by continuous lines, which are very well controlled at steady state reducing the uncertainty of current manufacturing and forecasting processes. Continuous does processes that enable close coupling of API and Drug Product operations. However, the need for buffers of intermediate materials should not be wholly discounted. From a quality perspective, control of particles at API should enable more reproducible drug substance manufacture. With the emerging technologies, 3D printing based supply models enabling local manufacture for a patient specific drug as part of future developments in personalized medicine.

IMPLEMENTATION OF CONTINUOUS MANUFACTURING

Control Systems

The consistency of product quality can be improved by using a combination of feedback and feed-forward control, and by monitoring slow variations such as catalyst deactivation over time at the plant-wide level so that adjustments can be sent to the control systems for the individual process unit operations level so that the final product leaving the last unit operation remains on-spec [11]. Plant-wide simulation allows the occasional application of cleaning or catalyst replacement protocols to be designed in such a way that product quality is not compromised, without having to shut down the entire plant.

An important consideration in the design of a continuous-flow process is the residence time distribution (RTD), which is the distribution function for the amount of time that a fluid element spends inside the vessel. Plug flow is one idealized RTD, in which there is no back-mixing and all fluid elements spend exactly the same amount of time in the vessel. The same notions apply to processes that have both fluids and particles. The reality is that there is always some intermixing of fluid elements, but many continuous-flow designs are specifically designed to approach plug flow as closely as possible. Plug flow operations create time delays in the process dynamics that can reduce the closed-loop performance achievable by the control system [11]. The effects of such time delays can often be mitigated by using feed-forward of measured disturbances to the control system or by having intermediate placement of sensors/analytics within the unit operation. Having such real-time detection, in addition to an understanding of RTD for the unit operation, affords the ability to track materials through the process to determine acceptability and the potential to isolate in the case of off-spec material.

One concern during continuous manufacturing is the modeling of variances, and how they might propagate through a continuous process. Specifically, if a disturbance enters the first unit operation in a series, the question arises as to whether that disturbance will spread out, and if so, by how much. To answer this question, an in depth understanding of the residence time distribution of each unit operation is required, as this serves as the transfer function which translates how input variability is related to variability in the output of each step. In addition to understanding each unit operation's transfer function, care must be taken to understand the system capacitance and delay times associated with material transport operations between steps. Because complex interconnected systems with time delays are inherently non-linear, care must be taken during design and testing of the individual and overall system controls to ensure that the system dynamics do not become unstable and lead to control system runaway or other chaotic phenomena.

Plant-wide simulation also enables the optimally sizing of the individual unit operations and recycle loops to maximize overall product yield and product quality of the manufacturing facility, and to track material as it moves through the plant. Any mathematical model allows the determination of the residence time distribution, and the largest time in which the value of the residence time distribution function is nonzero indicates the largest time in which material entering the process can stay in the process before exiting. This allows the tracking of any deviation through the process, to assess which products may be affected. The overall residence time distribution can be estimated by alternative methods such as carrying out scaling analyses or tracer experiments, but plant-wide simulation allows an analysis of the effects of changes in process operations on the residence time distribution.

Process Monitoring and Control

Continuous processes require control systems to ensure that the products are of high quality. Continuous processes differ from chemistries that involve only fluids, where substandard fluid can often be mixed with above-standard fluid to produce a fluid mixture that satisfies specifications, many pharmaceuticals processes require the manufacture of solids whose specifications cannot be met in this manner. For the pharmaceutical industry, the control systems must provide a higher assurance of consistent product quality than what is required in most processes in the chemical, oil refining, and petrochemical industries. Automation has the potential to improve the consistency of product quality by removing operator-induced variability, and much of the same process and product monitoring techniques used in the chemical, petrochemical, and oil refining industries can be applied to continuous pharmaceutical manufacturing. These process and product monitoring techniques include single- and multi-variable control charts, partial least squares, and principal component analysis [12]. Sensitivity analysis can be applied to the simulation model to determine bounds on the process variations in a single unit operation that ensure that the product quality leaving the last unit operation is within specifications. Many software packages such as offered by OSIsoft or AspenTech are available for archiving the large quantities of heterogeneous data collected in continuous processes that are used in process monitoring. Process analytical technology (PAT) instruments are well proven when used in a regulated environment; there is a high burden to provide proof of scientific understanding of the PAT method deployed [13]. Additionally, current interpretations of the 21 CFR Part 11 regulations are requiring manufacturers to keep historic records of all spectral data. This is a technical burden not experienced by other industries. Some of the most useful real-time online PAT data are spectroscopic [14]. The deployment of such sensors has been demonstrated in continuous pharmaceutical manufacturing. More broadly, it is desirable to have sufficient online PAT that the materials can be characterized to such a degree that offline analysis would be unnecessary. A key challenge in this regard is the ability to measure low levels (e.g., 0.15 wt% in final API) of often structurally similar impurities.

The in-process controls consist of analytical measurements, manipulated variables (e.g., pump flow rates), and the feedback controller. A common Quality-by-Design term is critical process parameters (CPP), which is the notion that a subset of process parameters are deemed critical for controlling the downstream product quality, with respect to the specified incoming material attributes or in process material attributes from an upstream process [13]. It is important to distinguish feed-forward control, in which manipulations are made in response to measurements of disturbances, from feed-back control, in which manipulations are made in response to measurements of variables that need to be controlled, such as compositions. A continuous process can be controlled in a truly dynamic fashion whereby downstream process parameters can be manipulated in response to measured upstream disturbances to maintain final product quality and in response to local real-time measurements of product quality. Using both feed-forward and feed-back control in response to measurements of disturbances throughout the process represents a valuable advantage to the continuous process. One approach to the design of a plant-wide control strategy is to develop and validate first-principles models for each unit operation, in isolation, and then place each of these unit operation models into a common platform for simulation of the entire manufacturing facility that is then used for the design and evaluation of a plant-wide control strategy [15].

Realistic models of uncertainties and disturbances were implemented in simulation to design and evaluate the effectiveness of the plant-wide control system before the construction of the continuous manufacturing facility was completed. Plant-wide simulations indicated that all product purity specifications would be satisfied once all of the control loops were closed, which was observed during the operation of the facility [1]. This application of model-based control design allows the reduction of the risk of running into unexpected operational or quality control problems once the manufacturing facility is brought online, and to evaluate whether the control system with the existing in-process process analytical technology/on-line monitoring can achieve real-time release instead of or in concert with end-product testing.

A system also needs to be in place to track material as it moves through the manufacturing facility, which is naturally handled by a combination of measurements and RTDs for the individual unit operations, and the equipment and process for segregating non-conforming material. This material traceability aspect is absolutely key for a sustainable pharmaceutical production facility, as the "area of influence" of a disturbance needs to be clearly understood and on a routine basis provides the basis to derive the data set of process variables that has been in place to transform the material stream in its way through the facility. Only profound knowledge of the system dynamics can be the basis for establishing the accurate and relevant description of the process conditions that have been in place for a particular unit dose of material.

System Integration

One view is that end-to-end, fully integrated continuous is the only way that should be considered. Anything else can be achieved with incremental steps to existing technology. This is perhaps a truism but represents the continuum between short term and longer-term vision in this arena. The control systems for the individual unit operations are coupled to a higher-level control layer, typically called the supervisory control layer. Supervisory control manages the flow rates as streams move through the manufacturing process, and manages impurities to ensure that any increases of impurities in upstream unit operations do not become so large that downstream processes cannot handle them, or have insufficient flexibility to deal with other disturbances. The supervisory control layer sends set points to the lower level controllers, with the main design criteria typically being to maintain product quality, ensure that there are no production rate mismatches, and that there are no operational problems due to recycle loops [16]. The lower level control loops usually employ real-time in-process measurements whereas supervisory control systems operate on slower time scales and can be updated with some analytical laboratory outputs.

Equipment and Process Tools

More recent developments have been focused on making continuous processing principles accessible to the industry by using familiar unit operations operated in a continuous fashion. A major enabler of continuous processing has been the commercialization of standalone laboratory bench top flow systems capable of performing chemistries under a broad range of temperatures and pressures. This availability has been matched by the provision of larger scaled processing units and the provision of off-the-shelf easily assembled components (including passive and active mixers, tubing unions, chemical resistant tubing), which can be assembled to create bespoke flow units. Comprehensive coverage of continuous reactor platforms is available detailing the operating principles and characteristics. Detailed understanding and characterization of the equipment should enable seamless scale-up with only minimal additional development. The essential requirement for realizing the benefits of flow is to ensure that robust control of each particular chemical and physical transformation involved is delivered by appropriate equipment whether that is bespoke or multipurpose.

If a continuous reactor is to be used in GMP production then several factors should be considered due to the continuous nature versus batch. This includes the scale of equipment to be used depending on whether the intent is to make a year's supply in one month or operate 24/7 and whether the plant is multipurpose or custom. In many cases a reaction can be run continuously but the feed solutions are prepared in batch and the workup and traditional batch processing also performs isolation. Startup and shutdown transition must be conducted in a way that minimizes or eliminates transition waste. If the product produced during the transition period still meets the required specifications then it can be forward processed even if the process was not truly at steady state during the transitions.

REGULATORY ASPECTS

Regulatory authorities in the three ICH regions and beyond are encouraging industry to adopt new technology as supported by ICH Q8(R2), Q9, Q10 and Q11, along with the introduction of Quality by Design (QbD) concepts, emphasizing science and risk based approaches to assure product quality [17, 18].

The emergence of ICH Q8 (R2), Q9, Q10, and Q11 guidelines and accompanying ICH Q-IWG Points to Consider (PTC) and Q and A documents emphasized that a prospective science and risk-based approach to development and lifecycle management could increase the assurance of quality of pharmaceutical products [5]. Collectively, these guidelines reinforced the adoption of risk–based (ICH Q9), systematic and science–based approaches (ICH Q8(R2) and ICH Q11), and a robust pharmaceutical quality system (ICH Q10), to establish an increased level of process understanding and product knowledge.

The *FDA Guidance for Industry PAT-A Framework for Innovative Pharmaceutical Development, Manufacturing, and Quality Assurance* specifically identifies that the introduction of continuous processing may be one of the outcomes from the adoption of a scientific risk-based approach to process design. Process understanding, control strategies, plus on-line, in-line, or at-line measurement of critical quality attributes provide for control strategies that include real time quality evaluation that is at least equivalent to, or better than, laboratory-based testing on collected samples.

EU Guidelines that might be particularly relevant to continuous manufacturing include the *Guidelines for Process Validation* where the concept of continuous process verification is introduced; the *Guideline on NIR* as it is often used as a Process Analytical Technology (PAT) tool for process monitoring and/or control, and the *Guideline on Real Time Release Testing*. Although not required, continuous manufacturing is commonly coupled with Real Time Release Testing (RTRT). Additionally, the European Medicines Agency set up a PAT Team in 2003 to support PAT and QbD activities in the EU. The teams act as a forum for dialogue between the Quality Working Party, the Biologics Working Party, and the Good Manufacturing Practice/Good Distribution Practice Inspectors' Working Group [19].

As the industry and Agencies become more familiar with continuous manufacturing, several regulatory considerations will need to be explored in order to link the principles and the practice. Within process validation and continuous process verification, process robustness and reproducibility should be evaluated [20]. The development of a continuous process should follow established principles, which are applied generally within pharmaceutical process development. Existing guidance and standards should also be consulted for process verification/validation and Continuous and/or Continued Process Verification.

Most recently, a presentation focusing specifically on continuous biomanufacturing was delivered by the FDA. The key conclusions were in line with previous guidance in the context of small molecule drugs. The main take home messages are that there is "nothing in regulations or guidance prohibiting continuous manufacturing," and "continuous manufacturing is fully consistent with FDA's QbD in which envelopes of performance are seen as part of the control strategy." The regulatory expectations for assurance of reliable and predictive processing in a commercial setting are the same for batch and continuous processing.

REFERENCES

[1] Mascia, S., Heider, P. L., Zhang, H., Lakerveld, R., Benyahia, B., Barton, P. I., Braatz, R. D., Cooney, C. L., Evans, J. M. B., Jamison, T. F., Jensen, K. F., Myerson, A. S. and Trout, B. L. (2013). End-to-end continuous manufacturing of pharmaceuticals: Integrated synthesis, purification, and final dosage formation. *Angew. Chem. Int. Ed.*, *52*(47), 12359-12363.

[2] Jayapal, K. R., Wlaschin, K. F., Hu, W. S. and Yap, M. G. S. (2007). Recombinant protein therapeutics from cho cells - 20 years and counting. *Chem. Eng. Prog.*, *103*(10), 40-47.

[3] Andersen, D. C. and Reilly, D. E. (2004). Production technologies for monoclonal antibodies and their fragments. *Curr. Opin. Biotechnol.*, *15*(5), 456-462.

[4] Li, F., Vijayasankaran, N., Shen, A., Kiss, R. and Amanullah, A. (2010). Cell culture processes for monoclonal antibody production. *mAbs.*, *2*(5), 466-477.

[5] Aksu, B., Beer, T. D., Folestad, S., Ketolainen, J., Lindén, H., Lopes, J. A., Matas, M. d., Oostra, W., Rantanen, J. and Weimer, M. (2012). Strategic funding priorities in the pharmaceutical sciences allied to quality by design (qbd) and process analytical technology (pat). *Eur. J. Pharm. Sci.*, *47*(2), 402-405.

[6] Schaber, S. D., Gerogiorgis, D. I., Ramachandran, R., Evans, J. M. B., Barton, P. I. and Trout, B. L. (2011). Economic analysis of integrated continuous and batch pharmaceutical manufacturing: A case study. *Ind. Eng. Chem. Res.*, *50*(17), 10083-10092.

[7] Roberge, D. M., Ducry, L., Bieler, N., Cretton, P. and Zimmermann, B. (2005). Microreactor technology: A revolution for the fine chemical and pharmaceutical industries? *Chem. Eng. Tech.*, *28*(3), 318-323.

[8] Glasnov, T. N. and Kappe, C. O. (2010). Toward a continuous-flow synthesis of boscalid®. *Adv. Synth. Catal.*, *352*(17), 3089-3097.

[9] Pennings, M., Pennings, M., Pennings, M., Pennings, M., Pennings, M. and Pennings, M. (2011). A continuous multicolumn chromatography process for polishing of mabs, in 7th HIC RPC Conference. Estoril, Portugal.

[10] Koutinas, M., Kiparissides, A., Pistikopoulos, E. N. and Mantalaris, A. (2012). Bioprocess systems engineering: Transferring traditional process engineering principles to industrial biotechnology. *Comput. Struct. Biotechnol. J.*, *3*(4), 1-9.

[11] Dietzsch, C., Spadiut, O. and Herwig, C. (2013). On-line multiple component analysis for efficient quantitative bioprocess development. *J. Biotechnol.*, *163*(4), 362-370.

[12] Simon, L. L., et al., (2015). Assessment of recent process analytical technology (pat) trends: A multiauthor review. *Org. Process Res. Dev.*, *19*(1), 3-62.

[13] Rathore, A. S. (2014). Qbd/pat for bioprocessing: Moving from theory to implementation. *Curr. Opin. Chem. Eng.*, *6*(0), 1-8.

[14] Wu, H., White, M. and Khan, M. A. (2011). Quality-by-design (qbd): An integrated process analytical technology (pat) approach for a dynamic pharmaceutical co-precipitation process characterization and process design space development. *Int. J. Pharm.*, *405*(1–2), 63-78.

[15] Poechlauer, P., Manley, J., Broxterman, R., Gregertsen, B. and Ridemark, M. (2012). Continuous processing in the manufacture of active pharmaceutical ingredients and finished dosage forms: An industry perspective. *Org. Process Res. Dev.*, *16*(10), 1586-1590.

[16] Lakerveld, R., Benyahia, B., Braatz, R. D. and Barton, P. I. (2013). Model-based design of a plant-wide control strategy for a continuous pharmaceutical plant. *AIChE J.*, *59*(10), 3671-3685.

[17] ICH Quality documents Q3, Q., Q8 Q8 (R2),Q9, Q10, Q11.

[18] PhRMA White Paper Implementation and Application of Quality by Design, Feb 2013.

[19] Woodcock, J. (2011). FDA, AAPS Annual Meeting October 2011.

[20] Chatterjee, S. (2012). FDA perspective on continuous manufacturing. IFPAC Meeting, January 2012.

In: High Value Processing Technologies
Editors: A. Berenjian, H. Jafarizadeh-Malmiri et al.
ISBN: 978-1-63484-453-6

Chapter 4

SUPERCRITICAL FLUID EXTRACTION OF VALUE ADDED COMPOUNDS FROM FOOD COMMODITIES

***Zahra Sayyar*[1], *Hoda Jafarizadeh-Malmiri*[1,*], *Shabnam Omrani*[1], *Aydin Berenjian*[2], *Hamideh Vaghari*[1] *and Omid Ahmadi*[1]**

[1]Faculty of Chemical Engineering,
Sahand University of Technology, Tabriz, Iran
[2]School of Engineering, Faculty of Science and Engineering,
The University of Waikato, Hamilton, New Zealand

ABSTRACT

Bioactive compounds of natural sources have unique sets of divers structures and functionalities which in turn, can be used as an excellent pool of molecules in fields ranging from food and mediciene to chemistry. The ever growing demand to extract plant bioactive compounds signifies the need of the most appropriate and standard method to extract these active components from plant materials. Supercritical fluid technology refers to any substance above its thermodynamic critical point. At this condition the diffusion coefficient is between those for liquids and gases and the density is close to that of liquids while viscosity is similar to that of gases. These unique properties are favourable in extraction and purification processes. The most common fluid, which has been broadly used in supercritical fluid extraction process, is carbon dioxide which has attracted much attention in food industries due to having moderate temperature and nontoxic nature. Supercritical carbon dioxide is widely used for extraction of essential oils from seeds, fruits, leaves, flowers, and rhizomes. This chapter tends to summarize the fundamentals of supercritical fluid extraction, their application in food industries and new directions in the research on natural matter separation.

[*] Corresponding author at: Faculty of Chemical Engineering, Sahand University of Technology, Tabriz, Iran, Tel: +98-413-3459099, Fax: +98-413-3444355, E-mail: h_jafarizadeh@sut.ac.ir; h_jafarizadeh@yahoo.com.

INTRODUCTION

Extraction of valuable bioactive compounds from natural products related to food and pharmaceutical is an important means of producing value added products. Phytochemical compounds (so-called bioactive components) have been studied for different functional activities including their antioxidant, antimicrobial, antihypertensive, anticancer, and anti-inflammatory properties. Application of separation technologies is used either to recover high value components from agricultural commodities, being an important operation for the production of food products such as oil and proteins, or especially for the development of health-promoting food ingredients and high value-added food products such as antioxidants and flavors. Antioxidant diet can be important for prevention or therapy of diseases [1-4]. Recently, demand for low calorie and low fat healthy food is increasing for diet nutrition. High protein-low calorie diet can be sourced from animal, such as whey protein, or from plant, such as seed from *Fabaceae* family or *Leguminosae* [3, 5]. Many potential high-value products can be developed from natural resources by different separation technologies and processes. Recently, there has been a lot of interest from the scientific community and increasing industrial demands to perform research and development on extraction and separation technologies based on eliminating the use of organic chemicals, environmental friendly aspects, minimum energy costs and human health regulations [6, 7]. Some conventional purification techniques, such as molecular separating, hydrodistillation and ion exchange chromatography are combined with other techniques to allow the achievement of high product purity, high efficiency and low energy costs without loss of product quality [1, 2]. For example, the essential oils from plants have been known to have potent antimicrobial and antioxidant activities with potential application in food products and pharmaceutical [8]. Extraction and purification of these compounds by hydrodistillation has been reported previously but the yields was no more than 0.5% [9]. Supercritical fluid extraction (SFE) is a new technology that may overcome to such problems. SFE is based on the solvating properties of supercritical fluid (SCF), which can be obtained by employing pressure and temperature above the critical point of a compound, mixture or element. SFE is regarded as a green process because it does not use chemical solvents with drastic environmental impacts. SFE have demonstrated that it can be more efficient alternative to conventional solvent extraction for essential oils and other bioactive components from plants such as flavour components, lycopene, β-carotene and lutein, because of the economics and environmental consequences of organic solvent usage and disposal [10-14]. Several studies have indicated that SFE could effectively remove oil content from high-protein diet source for producing healthy food with low calorie and fat content [5, 15]. Application of SFE as new separating method offers several advantages. For example, it does not employ fiery or toxic solvents and final products are without solvent residues with better release of aromatic compounds. Maximum separation speed, minimum effort, and minimum cost at as large a capacity as possible in production-scale processes, are other impressive advances of SFE in obtaining adequate segregations of components [6, 9, 16-18]. The aim of this review were to summarize and give an overview of the fundamentals of SFE with special attention to the type, properties and solubility parameter of SCF and co-solvent, and to provide a comprehensive review on SFE possible application in extraction of value added food bioactive components.

SOLVENT EXTRACTION

Solvent extraction is the removal of soluble components from solid or liquid materials by using an immiscible solvent. The process can be classified into solid-liquid and liquid-liquid extraction [17, 19-21]. Ideally, an analytical separation method should be rapid, simple and inexpensive; should give measurable recovery of analytes without loss or degradation. Separation of complex environmental, pharmaceutical, food and petroleum samples is mainly based upon extraction [20-23]. In this method, solvents should have the high distribution and diffusion coefficient, high selectivity for the solutes, and low miscibility with the feed. Furthermore, they should be stable and inert, nonflammable, no volatile, environmentally safe, and cheap [24, 25].

The most widely used methods for extraction of bioactive components from various plants and natural sources are based on hydrodistillation and solvent extraction. These methods are often associated with low yields and have the drawback of leaving trace amounts of organic solvents and cause thermal degradation of the more thermolabile compounds. In addition to, the latter uses solvents such as hexane are dangerous to handle and harmful to human health [26]. Extraction methods can have a significant impact on the yield, chemical composition and biological activities of bioactive compounds, particularly antioxidant and antimicrobial activities [8]. To meet food safety regulations, no toxic-chemical solvent residues are permitted in the end food products. The major issues related to bioactive compounds quality after separation are the impacts of processing on bioactivity of extracts and the nutritional value of the end products, as well as the quality characteristics. Therefore, food grade or pharmaceutical grade requirements cannot be achieved for added value phytochemical compounds using solvent extraction methods.

SUPERCRITICAL FLUID EXTRACTION

History and Definition

Supercritical fluid extraction is currently becoming alternative method to conventional extraction with solvent. A gas or a liquid is normally used as the extraction solvent for SFE. When a gas or liquid is compressed under pressure and heated past its critical point, it enters a phase called the supercritical phase and the extraction medium is called a supercritical fluid in this state [2, 17, 22]. SFE has been described as an environmentally safe technology due to compounds commonly extracted from natural source. SFE is a relatively new method in the field of analytical chemistry, having evolved in the last decade as an alternative method of preparing samples prior to analysis [16, 22, 27]. Intensive study of SFE in the 1970s and 1980s was generally confined to chemical processing applications that first used food industry such as the decaffeination of green coffee beans with supercritical carbon dioxide (SC-CO_2) [17]. SFE can be designed as a staged process for future applications that is the most effective and efficient way to extract valuable component of botanicals [22]. In the last years, the extracts of more than 300 plant species and natural materials have been studied using SFE technology [27]. In recent years, there has been an increasing interest in extraction of essential oils, antibacterial, antifungal, antioxidant, anti-inflammatory and pharmacological compounds from various herbs and aromatic plants using SFE due to its higher selectivity and recovery than conventional solvents [24, 28]. A simplified process-scale SFE system is shown in Figure 1. As clearly observed, raw material is charged in the extraction tank which is equipped with temperature controllers and pressure valves at both ends to keep desired extraction conditions. A supercritical fluid can be separated from analyte by simply releasing pressure [17, 22].

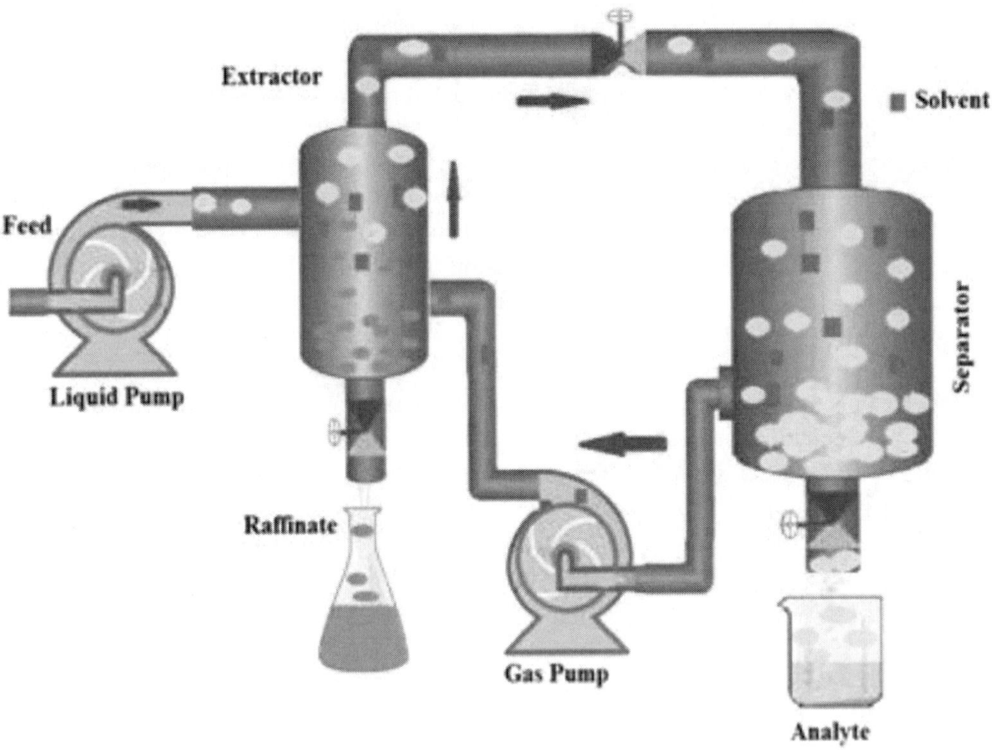

Figure 1. Schematic diagram of the supercritical fluid extraction.

Thermodynamic Properties of Supercritical Fluid

The thermodynamic state of a pure component is explained by three variables: pressure, temperature, and volume. Any substance is above its critical temperature and critical pressure is defined as a supercritical fluid (Figure 2) [29]. In addition, near the critical point, slight changes in pressure or temperature result in large changes in density [2, 17]. SCF undergo large reversible changes in their densities and correspondingly in their abilities to dissolve solutes near their critical points, and this unique property is the basis for their use in extractions and separations. SCFs are highly compressed gases, which have combined properties of gases and liquids in an intriguing manner [22]. A SCF exhibits physico-chemical properties intermediate between those of liquids and gases [2, 29]. It can penetrate into small pores within the matrix like a gas (low viscosity and surface tension), and its relatively high density gives good solvent power that dissolve materials like a liquid [30, 31]. The solvent power for low molecular weight compounds is high and decreases with increasing molecular weight [17, 32]. SCFs can lead to reactions, which are difficult or even impossible to achieve in conventional solvents. After extraction, SCF can be separated from analyte by simply releasing the pressure, leaving almost no trace and yields a pure residue [15, 22].

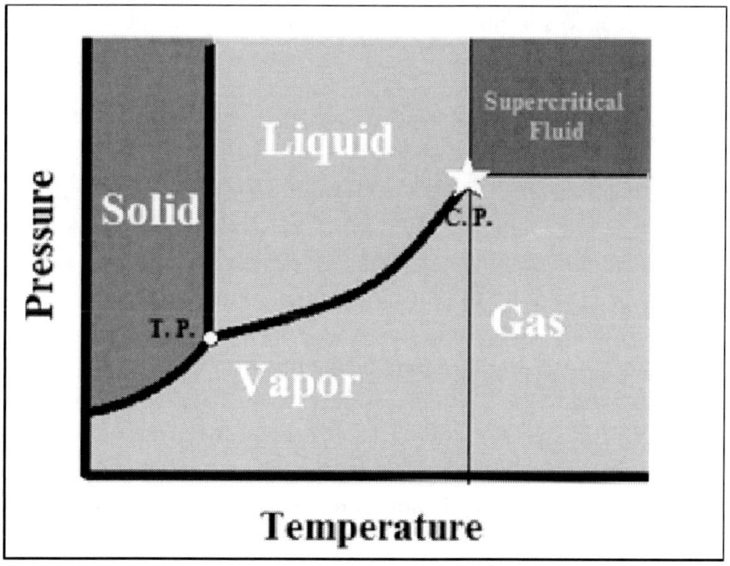

Figure 2. Phase diagram for a single substance, [T. P. (Triple Point), C. P. (Critical Point)].

Choice of Supercritical Fluid for Extraction

A number of SCFs are available with a range of values for density and critical temperatures. The choice of the solvent is noticed by the physicochemical properties of the compound of interest and role of solvent in the process in terms of solvency or non-solvency. Nonpolar materials have relatively low critical parameters (e.g., CO_2: T_c = 31.3, P_c = 72.9 atm); whereas, polar compounds have high critical parameters (e.g., NH_3: T_c = 132.5, P_c = 112.5 atm). Therefore, nonpolar materials allow supercritical operations at relatively low pressures and at near room temperatures which helps in preventing thermal degradation of food components being extracted [15-17]. Carbon dioxide, nitrogen oxide, ammonia, chlorofluroethane, trifluromathane, Xenon, nitrous oxide, sulphur hexafluoride and hydrocarbons such as hexane, pentane, butane and fluorinated hydrocarbons are the most commonly used SCFs in SFE technique [16, 19]. Among these fluids, CO_2 is the kind of extraction solvents for botanicals and bioactive compounds specially, essential oil due to its relatively inert nature, low cost, available in high purity, good solvent strength and miscibility with most solvent, non-toxicity and non-flammable [15-17, 33]. Furthermore, solute can be immediately separated from SC-CO_2 by reducing the pressure which in turn, pure solute obtain and there is no solvent residue in the extracted solutes [5, 15].

Choice of Co-Solvent

Generally, high extraction yield using SFE cannot be achieved using SCF, lonely. For example, CO_2 is not a very good solvent for high molecular weight and hydrophilic compounds [17]. For this reason, the use of other solvent compounds so-called co-solvent is needed in order to enhance solubility and the selectivity of the process and they must be added only in small quantities. In the case of CO_2 the important reason for adding a co-solvent to the supercritical phase relies on the affinity to polar solutes that modifies SC-CO_2 extraction efficiency [27]. To increase the solubility of such compounds in SC-CO_2 small amounts of polar or non-polar co-solvents (ranging from 0 to 20 mol%) which is also called modifiers may be added. In fact, the co-solvent interacts strongly with the solute and significantly increases the solubility [17, 34]. The application of co-solvents in SFE researches has been naturally dominated by ethanol, which is selected in most of the studies. Ethanol is a harmless solvent both to human health and environmental levels (like SC-CO_2) and this is a strong advantage comparing to n-hexane or even methanol, mainly when SFE is devoted to applications in food, or pharmaceutical industries. Besides the referred advantages, ethanol is significantly polar [27, 35]. One of the effects that co-solvent addition may have upon the sample matrix is swelling, which in turn may affect the cellular structure of the sample allowing a better access of the solvent to the matrix, and thus improving the diffusion of solute into the SC-CO_2 [2]. For example, the use of water as a co-solvent, meanwhile, improves lycopene extraction by swelling the sample matrix, increasing the mass transfer rate and barrier in the transport of the extract to the fluid [36, 37].

Important Factors in Supercritical Fluid Extraction Processes

The optimum extraction condition can be achieved by investigating multiple factors in all possible conditions [5]. While P and T affect the thermodynamic (density and solutes solubility) and transport properties (viscosity and diffusivity), solvent velocity is an independent variable that influences directly the axial dispersion, the convective mass transfer coefficient, and the accumulation in the bulk [27, 31, 38]. The extraction yield increases with the increase of the flow rate. By the increasing pressure at constant temperature, Effective diffusivity and mass transfer coefficient decrease but solubility, axial dispersion coefficient and extraction yield increase. In constant pressure, axial dispersion coefficient decreases while effective diffusivity, mass transfer coefficient and saturation concentration increases by the increasing temperature [39]. At a constant temperature, extraction at lower pressures will approval less polar analytes, and extraction at higher pressures will favor more polar and higher molecular weight analytes [40]. Some of the experimental factors must also be paid to the time period of extraction, fluid flow rate through the extraction cell, sample matrix effects, homogeneity of the sample, and system contamination. Sample Matrix Parameters that influence SFE such factors as particle size and shape, surface area and porosity, moisture content of the sample, Properties of solute (Solute properties, especially molecular weight, polarity, and vapor pressure), morphology and location of the solute in the matrix, and level of extractable matter will impact on the analytical results. In general, a decrease in particle size will improve the extraction efficiency and lead to a reduction in extraction time. For the extraction of nonpolar to moderately polar solutes in matrices containing high moisture levels, it is often necessary to remove most of the water by some technique [2, 7, 41]. In most cases, the water present in the vegetable matrix competes with the solute to interact with the solvent, decreasing the yield of the process. Also in some cases the presence of water is necessary to permit the interactions of the solvent with the solute, as in the extraction of caffeine from green coffee beans [42]. With regards to the above mentioned subject, solubility is important factor in the extraction in food industry, so we will discuss about it.

Solubility Parameter

SC-CO_2 density and solvent power of a compressed gas described by the Hildebrand solubility parameter which allows the establishment of a preliminary comparison of many solvents as long as the solute remains the same. The Hildebrand Solubility Parameter (δ) is a rough measure of the ability of a solvent to dissolve a solute [35], with the following dependence on the critical pressure, P_c, and reduced density at temperature T and at the normal boiling point, T_{eb}. It was calculated using following equation:

$$\delta = 1.25 p_c^{1/2} \left(\frac{\rho_r}{\rho_{r,eb}}\right) = 1.25 p_c^{1/2} \frac{\rho(P,T)}{\rho_{eb}(P,T_{eb})}$$

When the pressure is at atmospheric pressure, the solubility parameter is zero. The effect of pressure and temperature upon solubility can be briefly described: (i) pressure increases the SCF density, thus increasing its solvent power; (ii) temperature influences the solvent properties, specifically density, and also the solute properties, mostly vapor pressure. Increasing temperature, density decreases (lower solubility) and vapor pressure increases (higher solubility) so at low pressure, the density effect was dominant, while at higher pressures the solute vapor pressure was the leading mechanism affecting the extraction process [27, 35].

TYPES OF SUPERCRITICAL FLUID EXTRACTION

Supercritical fluid extraction allow for two extraction modes: one dynamic (continuous) and the other static (batch) (Figures 3 and 4). In both cases, the supercritical solvent is in contact with the material from which a desirable product is to be extracted. In the continuous mode, the extraction vessel is constantly supplied with fresh SCF, while the recovery is continually supplied with the solute. In the batch mode, the outlet of extraction vessel is closed and extraction takes place without the regeneration of SCF [43, 44]. SFE can be performed purely for sample preparation by collecting the extracted analytes for subsequent analysis by a variety of techniques such as "off-line" and "on-line" including chromatographic, spectroscopic, and gravimetric methods. The collection can be achieved either on-line into a chromatographic injection loop such as GC via an on-column injector, or off-line by depressurizing the SCF into a collection device that can be an empty vessel, a vessel containing a small volume of organic solvent, or solid-phase trap. Off-line SFE is inherently simpler than on-line SFE since the analyst needs only to consider the extraction and analyte collection steps [43]. Batch processes are preferable for small production of essential oil that may have to serve for different feeds and need to be highly flexible, whereas continuous processes are preferable for larger units dedicated to a specific production [45].

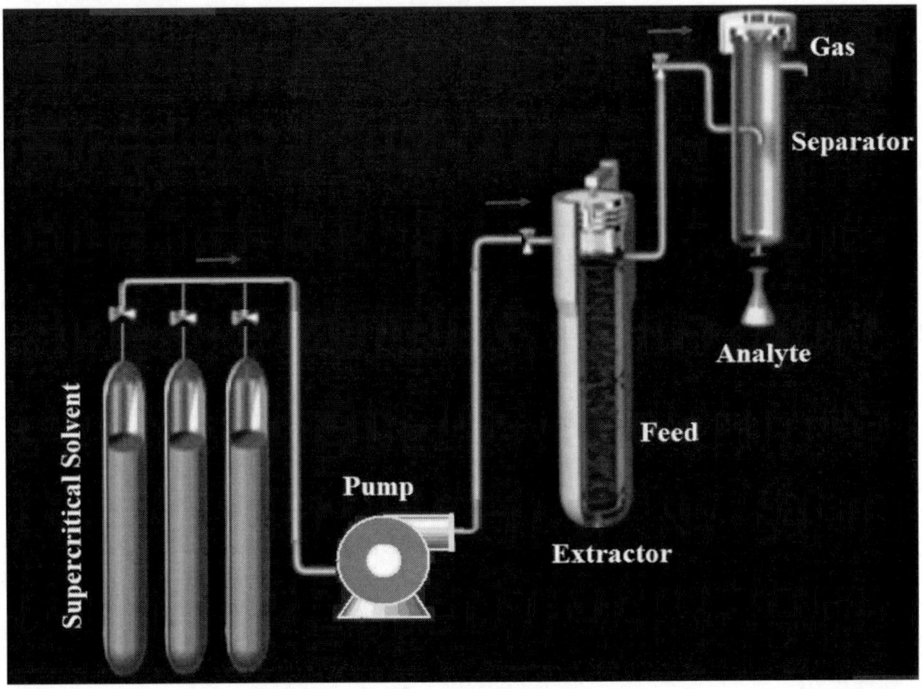

Figure 3. A schematic diagram of a supercritical fluid batch extraction.

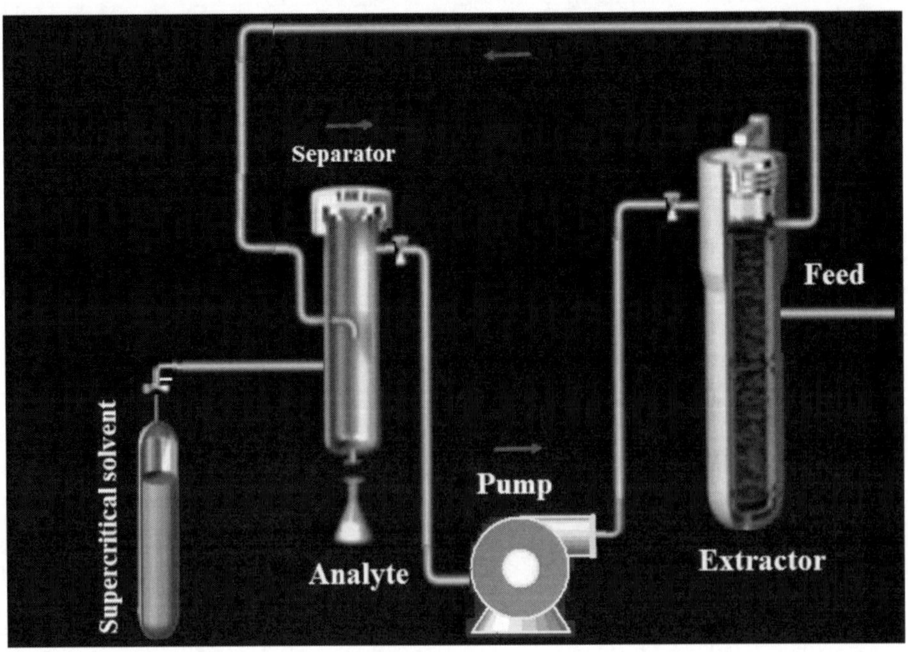

Figure 4. A schematic diagram of a supercritical fluid continuous extraction.

ADVANTAGES AND DISADVANTAGE OF SUPERCRITICAL FLUID EXTRACTION

Supercritical fluid extraction offers many advantages to the analyst that are not inherent in other sample preparation techniques, such as distillation, extraction with liquid solvents, or low resolution liquid chromatography [19, 23]. Indeed, SFE has immediate advantages over traditional extraction techniques. This system is a flexible process due to the possibility of continuous modulation of the solvent power/selectivity of the SCF [16, 34]. However, the base process scheme is relatively cheap and very simple to be scaled up to industrial scale [5, 16]. A comparison of SFE and solvent extraction is shown in Table 1 [5, 16, 19, 23, 28, 34, 40]. The biggest disadvantage of SFE is the cost associated with high-pressure equipment [5].

Table 1. Comparison of SFE with solvent extraction

Factors	Solvent Extraction	Supercritical Extraction
The amount of residual solvent	Solvent remain together with analyte. The residual amount of the solvent depends on the type of solvent used.	Extracted analytes are totally free of solvents and very pure, because elimination of solvents occur from them by releasing pressure.
Cost and yield	Solvent removal requires extra unit operations resulting in higher cost and lower recovery.	The reduction in operational steps, safe operation due to the use of nonorganic solvents. No extra unit operations needed and yield of useful material is very high.
Residual heavy metal	Heavy metal content is also unavoidable and depends on the solvent.	Totally free of heavy metals since there are no heavy metals in CO_2 or the equipment.
Residual inorganic salt	Inorganic salt content cannot be avoided, depends on the solvent.	Totally free of inorganic salts since there are no inorganic salts in CO_2 or the equipment.
Temperature condition	Not suitable for temperature-sensitive materials.	The use of moderate temperature in the critical range favorable for heat labile foods.
Selectivity	Polar substances get dissolved along with the lipophilic substances from the raw material due to poor selectivity of the solvent. All this causes the extract to look different from the basic components in the raw material and hence it is more of a "pseudo" natural extract.	No such possibility exists since CO_2 is highly selective and no chance of polar substances forming polymers exists.
Quality	Extracted analyte quality is not good.	The main advantage of SFE is the excellent quality of the resulting product.
Types of extracted	Both polar and non-polar are extracted	Only non-polar get extracted
Time	Liquid solvent extraction times can range from several hours to days.	SFE is fast quantitative that generally are complete in 10-60 min because Diffusivities are much faster.
Control of solvent strength	The solvent strength can hardly be controlled	The solvent strength of a SF can easily be controlled.

APPLICATIONS OF SUPERCRITICAL FLUID EXTRACTION

Bioprocessing can also be used in a variety of ways to produce value-added products. The various applications of biomaterials separation contain protein, enzyme, polysaccharide and flavor materials [46]. Some of applications of the SFE in food biotechnology can be summarized as follows: extraction of compounds from natural products (such as the extraction of caffeine, vanilla, β-carotene, vegetable oils, essential oil, antioxidant, etc.), removal of fat; removal of alcohol from wine and beer, encapsulation of liquids for engineering solid products, recovery of tocochromanols (vitamin E), i.e., enrichment of vitamin E from natural sources and beta-carotene (provitamin A) [15, 17, 22, 27, 47]. Furthermore, SFE extracts obtained from vegetable matrices are typically mixtures of the following family of compounds: triglycerides, fatty acids, fatty alcohols, terpenoids, phytosterols, tocopherols, tocotrienols, and phenolics that are presented in Figure 5 [27].

Table 2. Summary of publications comprising SFE of caffeine in food industry

Sample	P (MPa)	T (˚C)	Yield (%)	Ref.
Coffee	16–22	90	-	[48]
Tea	25–35	50–80	-	[49]
Green tea	20-40	40-80	54	[24]
Coffee	15.2–35.2	50–70	0–17	[50]
Tea stalk	20	55	-	[51]
Tea (green) Seed	5–9	35–45	14.8–29	[52]
Robusta coffee	30	100	84	[53]
Green tea leaves	23	63	96.6	[54]

Table 3. Summary of publications comprising SFE of cholesterol in food industry

Sample	P (MPa)	T (˚C)	Yield (%)	Ref.
Dried egg yolk	16.5-37.8	40-55	65.8	[55]
Dried chicken meat	30.6-37.6	45-55	90.0	[56]
Natural materials	31	40	-	[57]
Milk fat	10.1-36.4	40-70	91.5	[58]
Dried egg yolk	24.1-37.8	45-55	98.0	[59]
Egg	31	45	-	[60]
Butter Oil	10-27.6	40-70	-	[58]
Milk fat	8.0 -24.0	40-70	93.4	[61]
Black-eyed	25	60	5.40	[5]

SFE has been successfully used to extract oil from apricot, palm, canola, rape, soybean, sunflower, jojoba, sesame, celery, parsley, neem, amaranth, borage, flax and grape seeds. Oil has been also extracted from nuts such as acorn, walnut, almond and pistachio [28]. Some compilation of works in field of SFE in food industry for extraction of caffeine, cholesterol, fatty acid and antioxidants are available in Tables 2-5 respectively. Information regarding the species, extracted component and operating conditions (pressure, temperature, and yield) are provided for each SFE publication.

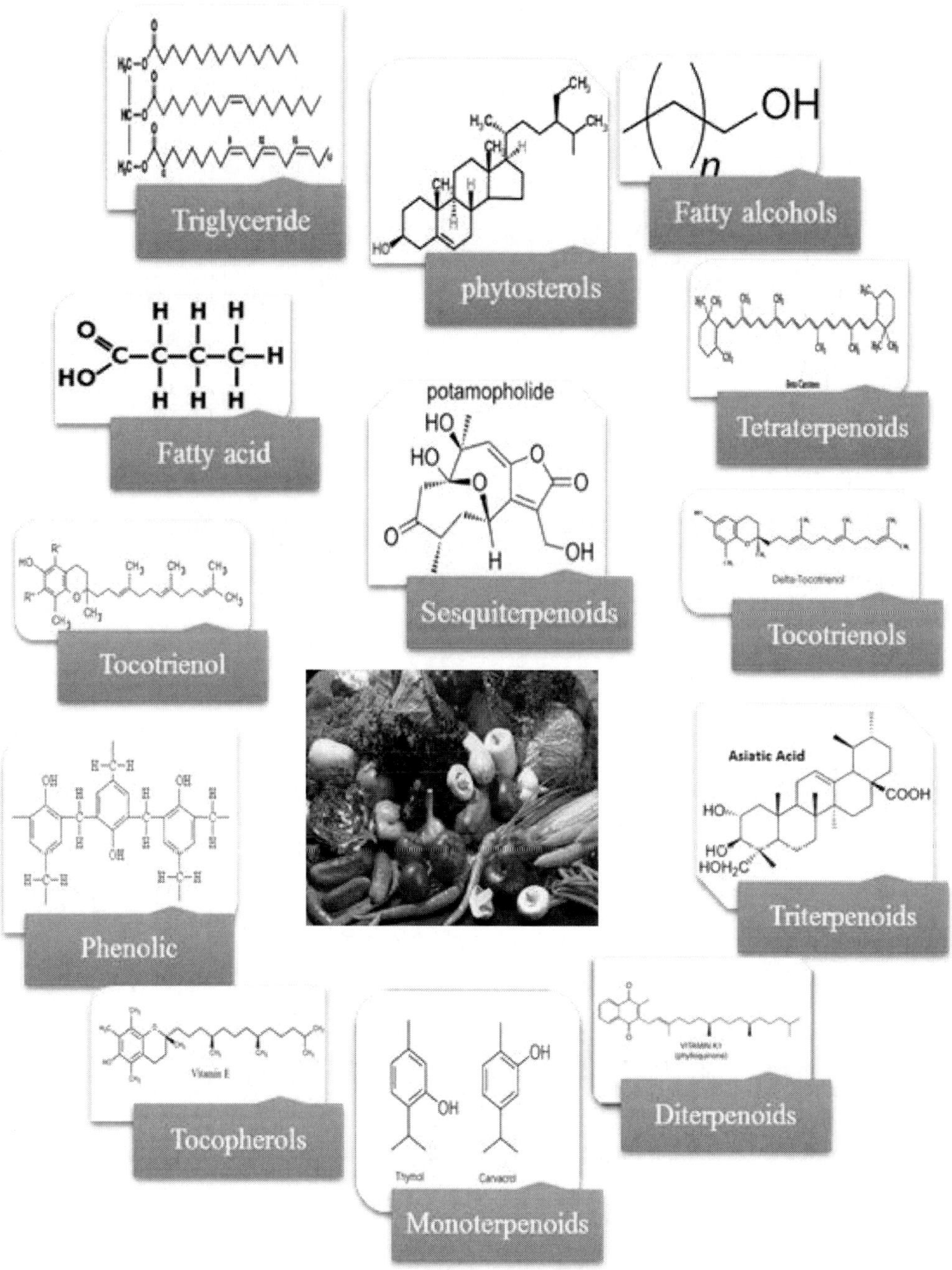

Figure 5. Main compounds found in the extracts obtained by SFE of vegetable biomass.

Table 4. Summary of publications comprising SFE of fatty acid in food industry

Sample	Analyte	P (MPa)	T (°C)	Yield (%)	Ref.
Milk	Fat	10–35	50–70	-	[62]
Rice brain	Fatty acid	17–31	1–60	22	[63]
Grapefruit	Limonene and Terpines	8-25	40	-	[64]
Black pepper	Essential oil	15–30	30–50	70	[10]
Coconut	Fatty acid	20.7-34.5	40–80	-	[65]
Sunflower	Fatty Acid	51.7-62	100	-	[66, 67]
Soybean	Fatty Acid	51.7-62	100	-	[66, 67]
Milk Fat	Lipid Profile Solid Fat content	6.9-17.2	40-60	-	[68-72]
Cherry Seeds	Essential oil	18–22	39–60	-	[73]
Grape seed	Oil	30-40	35-40	6.2	[74]
Rosmarinus officinalis	Essential oil	25	60	80	[11]
Grape seed	Fatty acid	-	-	13.6	[75]
Rosehip seed	Seed oil	15–45	40–80	15.93	[76]
Rose geranium	Essential oil	8–16	40–100	77.82	[77]
Cottonseed	Oil	35–55	60–80	17.26	[78]
Palm Kernel	Fatty acid	20.7–48.3	40–80	49	[12]
Wheat germ	Fatty acid	20–35	40–60	10.15	[79]
Cardamom seed	Fatty acid	10–30	35–55	-	[80]
Pistachio	Fatty acid	10–15	40–80	-	[81]
Sea buckthorn	Fatty acid	27.6	34.5	-	[82]
Pumpkin	Seed oil	15–35	35–75	30.7	[83]
Tomato and hazelnuts	Lycopene and essential oils	40	60	-	[84]
Sesame seed	Seed oil	19–25	40–60	35	[85]
Orange Pomace	Oleic acids n-Butyl benzenesulfonamide	10–30	40–50	0.85–3	[86]
Pithecellobium jiringan seed	Fatty acid,	60	80	-	[87]
Citrus peel	Essential oil	20	40-80	1.55	[37]
Cupressus sempervirens	Essential oil	90	40	34	[18]
Origanum glandulosum	Essential oil	90	40	2.1, 2.2	[4]
Black pepper	Oil	20	40	2.9	[33]
Rosemary	Essential oil	22	40	0.95-3.52	[13]
Peach seeds	Oil and β-sitosterol	16-24	35-55	35.3	[31]
Echium amoenum	Essential oil	15-30	40-55	-	[7]
Onion	Oil	20	35	0.53	[26]
Peanut	Glycerides of fatty acid short chain	35	60	48.53	[5]
Lavandula angustifolia	Essential oil	14.5	45	6.7	[6]
Juniperus communis	Essential oil	10-30	35-55	-	[14]
Milk thistle seeds	Oil and silybin	16-22	40-80	54.97	[38]
Marjoram and sweet basil	Essential oil	30	40	-	[88]

Table 5. Summary of publications comprising SFE of antioxidants in food industry

Sample	Analyte	P (MPa)	T (°C)	Yield (%)	Ref.
Carrot	Carotenes	60.6	40	-	[89]
Red grape skin	Anthocyanins	7–15	80	-	[90]
Grape Seeds	Polyphenolic compounds	45.6	35	-	[91]
Tomato paste waste	β-carotene and lycopene	20–30	35–65	-	[92]
Tomato skin	Lycopene	40.5	60–110	-	[93]
Tomato skin	Lycopene	40.5	60–110	-	[93]
Freeze-dried apricot bagasse	β-carotene	40.5, 30.4	40	-	[39]
Carrot Root	Carotenes	33	40	0.25	[94]
Hazel Fruit	β-Caryophyllene	30–45	40–60	1–17	[95]
Red pepper Fruit	Carotenoid	16.2–23	40	1–6	[96]
Tomato	Tocopherols and phytosterols	30	40	-	[97]
Tomato	Lycopene	25–45	40–70	-	[98]
Pine bark	Antioxidans	10–30	30–50	-	[99]
Rice wine	Lees Polyphenols	8	30	-	[100]
Fruit	Phenolic	8–30	60	1.4–5.20	[101]
Macela Flower	β-Asarone	20–30	30–35	87–96	[102]
Mangosteen Fruit	Xanthones	20–30	40–60	0.23–6.5	[103]

CONCLUSION

This chapter has illustrated some of the potential methods for using SFE to the analyst's advantage. SCFs exhibit many advantages for extraction in biotechnology processes such as high mass transfer rate, pressure sensitive solubility, and low temperature processing. SCFs extraction has also found several major applications to food processing include decaffeination of green coffee beans, production of hops extracts, recovery of aromas and flavors from herbs and spices, extraction and fractionation of seed oils and removal of contaminants. SFE can be used in a variety of modes for sample preparation and separation depending on the chosen product. This new technology may be mainly appropriate for difficult separations of high value products. In addition, SFE can provide fast, reliable, clean and cheap methods for routine analysis.

REFERENCES

[1] Khosravi-Darani, K. and Vasheghani-Farahani, E., "Application of supercritical fluid extraction in biotechnology," *Critical reviews in biotechnology,* vol. 25, pp. 231-242, 2005.

[2] Machmudah, S. and Goto, M., "Supercritical Fluid Extraction of Carotenoids," in *High Pressure Fluid Technology for Green Food Processing*, ed: Springer, 2015, pp. 397-426.

[3] Coelho, J. A. P. and Palavra, A. M. F., "Supercritical Fluid Extraction of Compounds from Spices and Herbs," in *High Pressure Fluid Technology for Green Food Processing*, ed: Springer, 2015, pp. 357-396.

[4] Mechergui, K., Jaouadi, W., Coelho, J. P., Serra, M. C., Marques, A. V., Palavra, A. M., et al., "Chemical composition and antioxidant activity of Tunisian Origanum glandulosum Desf. essential oil and volatile oil obtain by supercritical CO_2 extraction," *International Journal,* vol. 2, pp. 337-343, 2014.

[5] Anggrianto, K., Salea, R., Veriansyah, B., and Tjandrawinata, R. R., "Application of Supercritical Fluid Extraction on Food Processing: Black-eyed Pea (Vigna Unguiculata) and Peanut (Arachis Hypogaea)," *Procedia Chemistry,* vol. 9, pp. 265-272, 2014.

[6] Lesage-Meessen, L., Bou, M., Sigoillot, J.-C., Faulds, C. B., and Lomascolo, A., "Essential oils and distilled straws of lavender and lavandin: a review of current use and potential application in white biotechnology," *Applied microbiology and biotechnology,* vol. 99, pp. 3375-3385, 2015.

[7] Ghoreishi, S. M. and Bataghva, E., "Supercritical extraction of essential oil from Echium amoenum seed: Experimental, modeling and genetic algorithm parameter estimation," *Korean Journal of Chemical Engineering,* vol. 31, pp. 1632-1640, 2014.

[8] Danh, L. T., Triet, N. D. A., Zhao, J., Mammucari, R., and Foster, N., "Comparison of Chemical Composition, Antioxidant and Antimicrobial Activity of Lavender (Lavandula angustifolia L.) Essential Oils Extracted by Supercritical CO_2, Hexane and Hydrodistillation," *Food and Bioprocess Technology,* vol. 6, pp. 3481-3489, 2013.

[9] Wang, H., Liu, Y., Wei, S., and Yan, Z., "Application of response surface methodology to optimise supercritical carbon dioxide extraction of essential oil from Cyperus rotundus Linn," *Food Chemistry,* vol. 132, pp. 582-587, 2012.

[10] Ferreira, S. R., Nikolov, Z. L., Doraiswamy, L., Meireles, M. A. A., and Petenate, A. J., "Supercritical fluid extraction of black pepper (Piper nigrun L.) essential oil," *The Journal of Supercritical Fluids,* vol. 14, pp. 235-245, 1999.

[11] Santoyo, S., Cavero, S., Jaime, L., Ibanez, E., Senorans, F., and Reglero, G., "Chemical composition and antimicrobial activity of Rosmarinus officinalis L. essential oil obtained via supercritical fluid extraction," *Journal of Food Protection®,* vol. 68, pp. 790-795, 2005.

[12] Zaidul, I., Norulaini, N. N., Omar, A. M., and Smith, R., "Supercritical carbon dioxide (SC-CO_2) extraction of palm kernel oil from palm kernel," *Journal of Food Engineering,* vol. 79, pp. 1007-1014, 2007.

[13] Zermane, A., Larkeche, O., Meniai, A.-H., Crampon, C., and Badens, E., "Optimization of essential oil supercritical extraction from Algerian Myrtus communis L. leaves using response surface methodology," *The Journal of Supercritical Fluids,* vol. 85, pp. 89-94, 2014.

[14] Larkeche, O., Zermane, A., Meniai, A.-H., Crampon, C., and Badens, E., "Supercritical extraction of essential oil from Juniperus communis L. needles: Application of response surface methodology," *The Journal of Supercritical Fluids,* vol. 99, pp. 8-14, 2015.

[15] Darani, K. K. and Reza Mozafari, M., "Supercritical fluids technology in bioprocess industries: a review," *Journal of Biochemical Technology,* vol. 2, pp. 144-152, 2010.

[16] Reverchon, E. and De Marco, I., "Supercritical fluid extraction and fractionation of natural matter," *The Journal of Supercritical Fluids,* vol. 38, pp. 146-166, 2006.

[17] Abbas, K. A., Mohamed, A., Abdulamir, A., and Abas, H., "A review on supercritical fluid extraction as new analytical method," *American Journal of Biochemistry and Biotechnology,* vol. 4, pp. 345-353, 2008.

[18] Nejia, H., Severine, C., Jalloul, B., Mehrez, R., and Stephane, C. J., "Extraction of essential oil from Cupressus sempervirens: comparison of global yields, chemical composition and antioxidant activity obtained by hydrodistillation and supercritical extraction†," *Natural product research,* vol. 27, pp. 1795-1799, 2013.

[19] Herrero, M., Mendiola, J. A., Cifuentes, A., and Ibanez, E., "Supercritical fluid extraction: Recent advances and applications," *Journal of Chromatography A,* vol. 1217, pp. 2495-2511, 2010.

[20] Pereira, C. G. and Meireles, M. A. A., "Supercritical fluid extraction of bioactive compounds: fundamentals, applications and economic perspectives," *Food and Bioprocess Technology,* vol. 3, pp. 340-372, 2010.

[21] Akanda, M. J. H., Sarker, M. Z. I., Ferdosh, S., Manap, M. Y. A., Ab Rahman, N. N. N., and Ab Kadir, M. O., "Applications of supercritical fluid extraction (SFE) of palm oil and oil from natural sources," *Molecules,* vol. 17, pp. 1764-1794, 2012.

[22] Sapkale, G., Patil, S., Surwase, U., and Bhatbhage, P., " Supercritical fluid extraction," *Int. J. Chem. Sci.,* vol. 8(2), pp. 729-743 2010.

[23] Ibrahim, A., Al-Rawi, S., Majid, A. A., Rahman, N. A., Abo-Salah, K., and Ab Kadir, M., "Separation and Fractionation of Aquilaria Malaccensis Oil Using Supercritical Fluid Extraction and tThe Cytotoxic Properties of the Extracted Oil," *Procedia Food Science,* vol. 1, pp. 1953-1959, 2011.

[24] Kim, W.-J., Kim, J.-D., Kim, J., Oh, S.-G., and Lee, Y.-W., "Selective caffeine removal from green tea using supercritical carbon dioxide extraction," *Journal of Food Engineering,* vol. 89, pp. 303-309, 2008.

[25] Chemat, F., Fabiano-Tixier, A. S., Vian, M. A., Allaf, T., and Vorobiev, E., "Solvent-free extraction of food and natural products," *TrAC Trends in Analytical Chemistry,* 2015.

[26] Jin, J.-Z. and Tong, J.-Y., "Optimization of Supercritical Extraction of Oil from Onion by Supercritical CO_2," *Asian Journal of Chemistry,* vol. 26, p. 142, 2014.

[27] De Melo, M., Silvestre, A., and Silva, C., "Supercritical fluid extraction of vegetable matrices: applications, trends and future perspectives of a convincing green technology," *The Journal of Supercritical Fluids,* vol. 92, pp. 115-176, 2014.

[28] Xu, L., Zhan, X., Zeng, Z., Chen, R., Li, H., Xie, T., et al., "Recent advances on supercritical fluid extraction of essential oils," *African Journal of Pharmacy and Pharmacology,* vol. 5, pp. 1196-1211, 2011.

[29] Mchugh, M. and Krukonis, V., *Supercritical fluid extraction: principles and practice*: Elsevier, 2013.

[30] Hedrick, J. L., Mulcahey, L. J., and Taylor, L. T., "Supercritical fluid extraction," *Microchimica Acta,* vol. 108, pp. 115-132, 1992.

[31] Ekinci, M. S. and Guru, M., "Extraction of oil and β-sitosterol from peach (Prunus persica) seeds using supercritical carbon dioxide," *The Journal of Supercritical Fluids,* vol. 92, pp. 319-323, 2014.

[32] Sahena, F., Zaidul, I., Jinap, S., Karim, A., Abbas, K., Norulaini, N., et al., "Application of supercritical CO_2 in lipid extraction–a review," *Journal of Food Engineering,* vol. 95, pp. 240-253, 2009.

[33] Andrade, K. and Ferreira, S., "Antioxidant activity of black pepper (Piper nigrum L.) oil obtained by super critical CO_2," in *III Iberoamerican conference on super critical fluids, Cartagena de Indias (Colombia) pp*, 2013, pp. 1-5.

[34] Herrero, M., Cifuentes, A., and Ibanez, E., "Sub-and supercritical fluid extraction of functional ingredients from different natural sources: Plants, food-by-products, algae and microalgae: A review," *Food chemistry,* vol. 98, pp. 136-148, 2006.

[35] Rozzi, N. and Singh, R., "Supercritical fluids and the food industry," *Comprehensive reviews in food science and food safety,* vol. 1, pp. 33-44, 2002.

[36] Saldana, M. D., Temelli, F., Guigard, S. E., Tomberli, B., and Gray, C. G., "Apparent solubility of lycopene and β-carotene in supercritical CO_2, CO_2+ ethanol and CO_2+ canola oil using dynamic extraction of tomatoes," *Journal of Food Engineering,* vol. 99, pp. 1-8, 2010.

[37] Suetsugu, T., "Supercritical CO_2 extraction of essential oil from Kabosu (Citrus sphaerocarpa Tanaka) peel," *Flavour,* vol. 2, p. 18, 2013.

[38] Celik, H. T. and Guru, M., "Extraction of oil and silybin compounds from milk thistle seeds using supercritical carbon dioxide," *The Journal of Supercritical Fluids,* vol. 100, pp. 105-109, 2015.

[39] Doker, O., Salgın, U., Sanal, I., Mehmetoğlu, U., and Çalımlı, A., "Modeling of extraction of β-carotene from apricot bagasse using supercritical CO_2 in packed bed extractor," *The Journal of supercritical fluids,* vol. 28, pp. 11-19, 2004.

[40] Lee, M. L. and Markides, K. E., "Analytical supercritical fluid chromatography and extraction," 1990.

[41] Sairam, P., Ghosh, S., Jena, S., Rao, K., and Banji, D., "Supercritical Fluid Extraction (SFE)-An Overview," *AJPSci,* vol. 2, pp. 112-20, 2012.

[42] Lehotay, S. and Ibrahim, M., "Supercritical fluid extraction and gas chromatography/ion trap mass spectrometry of pentachloronitrobenzene pesticides in vegetables," *Journal of AOAC International,* vol. 78, pp. 445-452, 1994.

[43] Vyasn, K. M., Panchal, S., Butani, A., and Kumar, V., "Supercritical fluid technology-An unlimited frontier in herbal research," *Int J Pharm Sci Nanotechnol,* vol. 1, pp. 303-307, 2009.

[44] Mohamed, R. S. and Mansoori, G. A., "The use of supercritical fluid extraction technology in food processing," *Food Technology Magazine,* vol. 20, pp. 134-139, 2002.

[45] Gironi, F. and Maschietti, M., "Continuous countercurrent deterpenation of lemon essential oil by means of supercritical carbon dioxide: Experimental data and process modelling," *Chemical Engineering Science,* vol. 63, pp. 651-661, 2008.

[46] King, J. W., "Analytical-process supercritical fluid extraction: a synergestic combination for solving analytical and laboratory scale problems," *TrAC Trends in Analytical Chemistry,* vol. 14, pp. 474-481, 1995.

[47] del Valle, J. M., "Extraction of natural compounds using supercritical CO_2: Going from the laboratory to the industrial application," *The Journal of Supercritical Fluids,* vol. 96, pp. 180-199, 2015.

[48] Williams, D., "Extraction with supercritical gases," *Chemical Engineering Science,* vol. 36, pp. 1769-1788, 1981.

[49] Calabuig Aracil, J., "Extraccion con CO_2 supercrítico de la cafeina del cafe y otras aplicaciones alimentarias," *Trabajo Final de Carrera. Departamento de Industrias Agroalimentarias. Escola Superior d'Agricultura de Barcelona. UPC,* 1998.

[50] De Azevedo, A., Mazzafera, P., Mohamed, R., Melo, S., and Kieckbusch, T., "Extraction of caffeine, chlorogenic acids and lipids from green coffee beans using supercritical carbon dioxide and co-solvents," *Brazilian Journal of Chemical Engineering,* vol. 25, pp. 543-552, 2008.

[51] Icen, H. and Gürü, M., "Extraction of caffeine from tea stalk and fiber wastes using supercritical carbon dioxide," *The Journal of Supercritical Fluids,* vol. 50, pp. 225-228, 2009.

[52] Wang, Y., Sun, D., Chen, H., Qian, L., and Xu, P., "Fatty acid composition and antioxidant activity of tea (Camellia sinensis L.) seed oil extracted by optimized supercritical carbon dioxide," *International journal of molecular sciences,* vol. 12, pp. 7708-7719, 2011.

[53] Tello, J., Viguera, M., and Calvo, L., "Extraction of caffeine from Robusta coffee (Coffea canephora var. Robusta) husks using supercritical carbon dioxide," *The Journal of Supercritical Fluids,* vol. 59 (11), pp. 53-60, 2011.

[54] Park, H. S., Im, N. G., and Kim, K. H., "Extraction behaviors of caffeine and chlorophylls in supercritical decaffeination of green tea leaves," *LWT - Food Science and Technology,* vol. 45 (1), pp. 73-78, 2012.

[55] Froning, G., Wehling, R., Cuppett, S., Pierce, M., Niemann, L., and Siekman, D., "Extraction of cholesterol and other lipids from dried egg yolk using supercritical carbon dioxide," *Journal of Food Science,* vol. 55, pp. 95-98, 1990.

[56] Froning, G., Fieman, F., Wehling, R., Cuppett, S., and Niemann, L., "Supercritical carbon dioxide extraction of lipids and cholesterol from dehydrated chicken meat," *Poultry science,* vol. 73, pp. 571-575, 1994.

[57] Nguyen, U., Evans, D., and Frakman, G., "Natural antioxidants produced by supercritical extraction," *Supercritical fluid processing of food and biomaterials,* p. 103, 1994.

[58] Mohamed, R. S., Neves, G., and Kieckbusch, T. G., "Reduction in cholesterol and fractionation of butter oil using supercritical CO_2 with adsorption on alumina," *International journal of food science and technology,* vol. 33, pp. 445-454, 1998.

[59] Bohac, C., Rhee, K., Cross, H., and Ono, K., "Assessment of methodologies for calorimetric cholesterol assay of meats," *Journal of Food Science,* vol. 53, pp. 1642-1644, 1988.

[60] Froning, G., Wehling, R., Cuppett, S., and Niemann, L., "Moisture content and particle size of dehydrated egg yolk affect lipid and cholesterol extraction using supercritical carbon dioxide," *Poultry science,* vol. 77, pp. 1718-1722, 1998.

[61] Mohamed, R. S., Saldana, M. D., Socantaype, F. H., and Kieckbusch, T. G., "Reduction in the cholesterol content of butter oil using supercritical ethane extraction and adsorption on alumina," *The Journal of Supercritical Fluids,* vol. 16, pp. 225-233, 2000.

[62] ARUL, J., Boudreau, A., Makhlouf, J., Tardif, R., and Sahasrabudhe, M. R., "Fractionation of anhydrous milk fat by superficial carbon dioxide," *Journal of Food Science,* vol. 52, pp. 1231-1236, 1987.

[63] Shen, Z., Palmer, M. V., Ting, S. S., and Fairclough, R. J., "Pilot scale extraction and fractionation of rice bran oil using supercritical carbon dioxide," *Journal of Agricultural and Food Chemistry,* vol. 45, pp. 4540-4544, 1997.

[64] Poiana, M., Sicari, V., and Mincione, B., "Supercritical carbon dioxide (SC-CO_2) extraction of grapefruit flavedo," *Flavour and fragrance journal,* vol. 13, pp. 125-130, 1998.

[65] Mira, B., Blasco, M., Berna, A., and Subirats, S., "Supercritical CO_2 extraction of essential oil from orange peel. Effect of operation conditions on the extract composition," *The Journal of supercritical fluids,* vol. 14, pp. 95-104, 1999.

[66] Taylor, S. L., Eller, F. J., and King, J. W., "A comparison of oil and fat content in oilseeds and ground beef—using supercritical fluid extraction and related analytical techniques," *Food research international,* vol. 30, pp. 365-370, 1997.

[67] Brühl, L. and Matthäus, B., "Extraction of oilseeds by SFE–a comparison with other methods for the determination of the oil content," *Fresenius' journal of analytical chemistry,* vol. 364, pp. 631-634, 1999.

[68] Bhaskar, A., Rizvi, S., Bertoli, C., Fay, L., and Hug, B., "A comparison of physical and chemical properties of milk fat fractions obtained by two processing technologies," *Journal of the American Oil Chemists' Society,* vol. 75, pp. 1249-1264, 1998.

[69] Berg, H., Turner, C., Dahlberg, L., and Mathiasson, L., "Determination of food constituents based on SFE: applications to vitamins A and E in meat and milk," *Journal of biochemical and biophysical methods,* vol. 43, pp. 391-401, 2000.

[70] Manganiello, L., Ríos, A., and Valcarcel, M., "Automatic microgravimetric determination of fats in milk products by use of supercritical fluid extraction with on-line piezoelectric detection," *Journal of Chromatography A,* vol. 874, pp. 265-274, 2000.

[71] Ramos, L., Hernandez, L., and Gonzalez, M., "Study of the distribution of the polychlorinated biphenyls in the milk fat globule by supercritical fluid extraction," *Chemosphere,* vol. 41, pp. 881-888, 2000.

[72] Turner, C. and Mathiasson, L., "Determination of vitamins A and E in milk powder using supercritical fluid extraction for sample clean-up," *Journal of Chromatography A,* vol. 874, pp. 275-283, 2000.

[73] Bernardo-Gil, G., Oneto, C., Antunes, P., Rodrigues, M. F., and Empis, J. M., "Extraction of lipids from cherry seed oil using supercritical carbon dioxide," *European Food Research and Technology,* vol. 212, pp. 170-174, 2001.

[74] Cao, X. and Ito, Y., "Supercritical fluid extraction of grape seed oil and subsequent separation of free fatty acids by high-speed counter-current chromatography," *Journal of Chromatography A,* vol. 1021, pp. 117-124, 12/22/ 2003.

[75] Beveridge, T. H., Girard, B., Kopp, T., and Drover, J. C., "Yield and composition of grape seed oils extracted by supercritical carbon dioxide and petroleum ether: varietal effects," *Journal of agricultural and food chemistry,* vol. 53, pp. 1799-1804, 2005.

[76] Machmudah, S., Kawahito, Y., Sasaki, M., and Goto, M., "Supercritical CO 2 extraction of rosehip seed oil: fatty acids composition and process optimization," *The Journal of supercritical fluids,* vol. 41, pp. 421-428, 2007.

[77] Gomes, P. B., Mata, V. G., and Rodrigues, A. E., "Production of rose geranium oil using supercritical fluid extraction," *The Journal of Supercritical Fluids,* vol. 41, pp. 50-60, 2007.

[78] Bhattacharjee, P., Singhal, R. S., and Tiwari, S. R., "Supercritical carbon dioxide extraction of cottonseed oil," *Journal of Food Engineering,* vol. 79, pp. 892-898, 2007.

[79] Shao, P., Sun, P., and Ying, Y., "Response surface optimization of wheat germ oil yield by supercritical carbon dioxide extraction," *Food and Bioproducts Processing,* vol. 86, pp. 227-231, 2008.

[80] Hamdan, S., Daood, H. G., Toth-Markus, M., and Illés, V., "Extraction of cardamom oil by supercritical carbon dioxide and sub-critical propane," *The Journal of Supercritical Fluids,* vol. 44, pp. 25-30, 2008.

[81] Sheibani, A. and Ghaziaskar, H. S., "Pressurized fluid extraction of pistachio oil using a modified supercritical fluid extractor and factorial design for optimization," *Lwt-Food Science and Technology,* vol. 41, pp. 1472-1477, 2008.

[82] Xu, X., Gao, Y., Liu, G., Wang, Q., and Zhao, J., "Optimization of supercritical carbon dioxide extraction of sea buckthorn (Hippophae thamnoides L.) oil using response surface methodology," *LWT-Food Science and Technology,* vol. 41, pp. 1223-1231, 2008.

[83] Mitra, P., Ramaswamy, H. S., and Chang, K. S., "Pumpkin (Cucurbita maxima) seed oil extraction using supercritical carbon dioxide and physicochemical properties of the oil," *Journal of food engineering,* vol. 95, pp. 208-213, 2009.

[84] Ciurlia, L., Bleve, M., and Rescio, L., "Supercritical carbon dioxide co-extraction of tomatoes (Lycopersicum esculentum L.) and hazelnuts (Corylus avellana L.): A new procedure in obtaining a source of natural lycopene," *The Journal of Supercritical Fluids,* vol. 49, pp. 338-344, 2009.

[85] Corso, M. P., Fagundes-Klen, M. R., Silva, E. A., Cardozo Filho, L., Santos, J. N., Freitas, L. S., et al., "Extraction of sesame seed (Sesamun indicum L.) oil using compressed propane and supercritical carbon dioxide," *The Journal of Supercritical Fluids,* vol. 52, pp. 56-61, 2010.

[86] Benelli, P., Riehl, C. A., Smânia, A., Smânia, E. F., and Ferreira, S. R., "Bioactive extracts of orange (Citrus sinensis L. Osbeck) pomace obtained by SFE and low pressure techniques: Mathematical modeling and extract composition," *The Journal of Supercritical Fluids,* vol. 55, pp. 132-141, 2010.

[87] Norulaini, N. N., Zaidul, I., Azizi, C. M., Zhari, I., Noramin, M., Sahena, F., et al., "Supercritical carbon dioxide fractionation of pithecellobium jiringan jack seed compositions using fast gas chromatography time of flight mass spectrometry," *Journal of Food Process Engineering,* vol. 34, pp. 1746-1758, 2011.

[88] Arranz, E., Jaime, L., de las Hazas, M. L., Reglero, G., and Santoyo, S., "Supercritical fluid extraction as an alternative process to obtain essential oils with anti-inflammatory properties from marjoram and sweet basil," *Industrial Crops and Products,* vol. 67, pp. 121-129, 2015.

[89] Chandra, A. and Nair, M. G., "Supercritical fluid carbon dioxide extraction of α-and β-carotene from carrot (Daucus carota L.)," *Phytochemical Analysis,* vol. 8, pp. 244-246, 1997.

[90] Blasco, M., Tarrega, A., Capilla, V., and Subirats, S., "Applications of SCF in food industry," *AINIA, Valencia, Spain,* 1999.

[91] Palma, M., Taylor, L. T., Varela, R. M., Cutler, S. J., and Cutler, H. G., "Fractional extraction of compounds from grape seeds by supercritical fluid extraction and analysis for antimicrobial and agrochemical activities," *Journal of agricultural and food chemistry,* vol. 47, pp. 5044-5048, 1999.

[92] Baysal, T., Ersus, S., and Starmans, D., "Supercritical CO_2 extraction of β-carotene and lycopene from tomato paste waste," *Journal of Agricultural and Food Chemistry,* vol. 48, pp. 5507-5511, 2000.

[93] Ollanketo, M., Hartonen, K., Riekkola, M.-L., Holm, Y., and Hiltunen, R., "Supercritical carbon dioxide extraction of lycopene in tomato skins," *European Food Research and Technology,* vol. 212, pp. 561-565, 2001.

[94] Ranalli, A., Contento, S., Lucera, L., Pavone, G., Di Giacomo, G., Aloisio, L., et al., "Characterization of carrot root oil arising from supercritical fluid carbon dioxide extraction," *Journal of agricultural and food chemistry,* vol. 52, pp. 4795-4801, 2004.

[95] Ozkal, S., Yener, M., Salgın, U., and Mehmetoglu, U., "Response surfaces of hazelnut oil yield in supercritical carbon dioxide," *European Food Research and Technology,* vol. 220, pp. 74-78, 2005.

[96] Gouveia, A. F., Duarte, C., Beirao da Costa, M. L., Bernardo-Gil, M. G., and Moldao-Martins, M., "Oxidative stability of olive oil flavoured by Capsicum frutescens supercritical fluid extracts," *European journal of lipid science and technology,* vol. 108, pp. 421-428, 2006.

[97] Vagi, E., Simandi, B., Vasarhelyine, K., Daood, H., Kéry, A., Doleschall, F., et al., "Supercritical carbon dioxide extraction of carotenoids, tocopherols and sitosterols from industrial tomato by-products," *The Journal of supercritical fluids,* vol. 40, pp. 218-226, 2007.

[98] Kassama, L. S., Shi, J., and Mittal, G. S., "Optimization of supercritical fluid extraction of lycopene from tomato skin with central composite rotatable design model," *Separation and Purification Technology,* vol. 60, pp. 278-284, 2008.

[99] Braga, M. E., Santos, R. M., Seabra, I. J., Facanali, R., Marques, M. O., and de Sousa, H. C., "Fractioned SFE of antioxidants from maritime pine bark," *The Journal of Supercritical Fluids,* vol. 47, pp. 37-48, 2008.

[100] Wu, J.-J., Lin, J.-C., Wang, C.-H., Jong, T.-T., Yang, H.-L., Hsu, S.-L., et al., "Extraction of antioxidative compounds from wine lees using supercritical fluids and associated anti-tyrosinase activity," *The Journal of Supercritical Fluids,* vol. 50, pp. 33-41, 2009.

[101] Laroze, L. E., Díaz-Reinoso, B., Moure, A., Zúñiga, M. E., and Domínguez, H., "Extraction of antioxidants from several berries pressing wastes using conventional and supercritical solvents," *European Food Research and Technology,* vol. 231, pp. 669-677, 2010.

[102] Hatami, T., Cavalcanti, R. N., Takeuchi, T. M., and Meireles, M. A. A., "Supercritical fluid extraction of bioactive compounds from Macela (Achyrocline satureioides) flowers: Kinetic, experiments and modeling," *The Journal of Supercritical Fluids,* vol. 65, pp. 71-77, 2012.

[103] Zarena, A., Manohar, B., and Sankar, K. U., "Optimization of supercritical carbon dioxide extraction of xanthones from mangosteen pericarp by response surface methodology," *Food and Bioprocess Technology,* vol. 5, pp. 1181-1188, 2012.

In: High Value Processing Technologies
Editors: A. Berenjian, H. Jafarizadeh-Malmiri et al.
ISBN: 978-1-63484-453-6

Chapter 5

STRUCTURE, SYNTHESIS AND ENVIRONMENTAL APPLICATIONS OF NANO-SIZED HYDROXYAPATITE

Shiva Salem[1], Amin Salem[2,3,*], Mahboube Sadevandi[3] and Elmira Velayi[4]

[1]Faculty of Chemical Engineering, Urmia University of Technology, Urmia, Iran
[2]Mineral Processing Research Center, Chemical Engineering Department, Sahand University of Technology, Tabriz, Iran
[3]Center of Excellence for Color Science and Technology, Tehran, Iran
[4]School of Chemical Engineering, Iran University of Science and Technology, Tehran, Iran

ABSTRACT

Hydroxyapatite is a calcium phosphate based material which is well known for the environmental applications due to excellent adsorption capacity for the removal of contaminants from water and wastewaters. Recently, the nano-scale hydroxyapatite has been found the developed employment in the environmental treatments. The purpose of present chapter is to organize the scattered available information on various aspects regarding structure, synthesis and applications of nano-sized hydroxyapatite as a potentially advance adsorbent for the recovery of wastewaters. It is evident that the synthesis factors significantly affect the structure and morphology of final products therefore, the outstanding conditions which improve the adsorption capability have demonstrated.

Keywords: hydroxyapatite, nano-particles, structure, synthesis, wastewater, recovery

INTRODUCTION

The presence of contaminates such as heavy metal ions and dyes in the effluents is a major concern due to threats of these toxic materials for humans and other living beings.

---

* Corresponding Author Email Address: salem@sut.ac.ir.

Natural water may be contaminated with several heavy metal ions arising from mining wastes and industrial discharges such as chemical, metallurgical, electrical and electronic units. Also, the textile, tannery, paper, rubber, plastic and paint inevitably make the flow of organic substances such as cationic dyes into the environment. Although, the cationic dyes such as methylene blue are visible in wastewater, the heavy metal ions are commonly invisible substances. The treatment of effluents containing contaminates is of interest due to harmful effects on water, ground water and soil. The several physical, chemical and biological techniques have been reported for the removal of contaminates below the maximal permissible level before contact with water and soil. Amongst the numerous techniques for removing metal cations [1, 2] and dyes [3-5], adsorption is the applicable choice and gives the best results due to low initial cost, flexibility and simplicity of design, ease of operation and insensitivity to toxic pollutants.

Many conventional adsorbents including activated carbon [6], zeolite [7-9] and clays [10-14] have been employed for uptake of contaminates. However, the advance materials such as micro and nano-sized hydroxyapatite (HA) can be used as efficient adsorbents for recovery of wastewaters. In fact, the nano-sized HA offers the novel efficiency in comparison to micro-scale particles. Several methods have been applied for the synthesis of HA in nano-scale. The adsorption on HA is a well known equilibrium separation process and an effective method for water decontamination [15, 16]. The present chapter deals the structure, synthesis and application of nano-sized HA for recovery of wastewaters. The main aim is to provide a summary of information, concerning the synthesis factors and adsorption capacity. In order to achieve this goal, extensive reported results in literature have been compiled. The reader is strongly recommended to study the original articles for detail information on experimental conditions.

HYDROXYAPATITE STRUCTURE

Calcium phosphate compositions are of great importance in different fields of chemical engineering such as biomaterials and recovery processes [17]. HA is the most stable calcium phosphate salt at normal environmental conditions. The chemical formula of HA is $Ca_{(10-x)}(HPO_4)_x(PO_4)_{(6-x)}(OH)_{(2-x)}$ where x varies between 0 and 1, depending on specifications [18]. By considering the minimal and maximal amounts of calcium, the HA formula may be written as $Ca_9HPO_4(PO_4)_5OH$ which is known as calcium-deficient HA and or $Ca_{10}(PO_4)_6(OH)_2$ as stoichiometric HA. The HA lattice is constituted by Ca^{2+}, PO_4^{2-} and OH^- which are arranged in a hexagonal cell [19]. Although it is very difficult to avoid the substitution of other ions in the lattice, a phase close to HA formula can be synthesized from solutions containing calcium and phosphate ions. This phenomenon can be attributed to the hexagonal structure, constructed from columns of divalent metal ions and oxygen atoms forming parallel channels in the lattice, making it particularly prone to ionic substitutions. The thermal decomposition of HA depends on Ca/P ratio [20-22]. Tri-calcium phosphate, $Ca_3(PO_4)_2$, appears at higher temperatures with an increase in Ca/P ratio. Calcium oxide (CaO) is also found when Ca/P > 1.7. The most thermally stable HA is prepared with nearly stoichiometric mixing ratios from 1.68 to 1.70. In comparison to stoichiometric HA, the thermal decomposition easily occurs for calcium-deficient HA.

A wide variety of substitutions can be made in the HA lattice by large number of anions, cations, and functional groups. There are two distinct calcium sites in the HA lattice where substitutions can occur by metal ions, such as iron. Using a ferric chloride solution and a simple soaking procedure, it is possible to prepare Fe-HA with no apparent formation of a second phase [23]. This procedure is unsuccessful when ferrous chloride solution is used. The substitution of Fe^{3+} into the HA lattice results in Fe-HA powders with magnetic properties. This procedure can be applied to synthesize magnetic HA nanoparticles for the environmental applications.

SYNTHESIS TECHNOLOGIES

The frequently used wet chemical methods for production of HA include precipitation, hydrothermal process, sol-gel synthesis, emulsion processing and autoignition technique. The wet chemical reactions for HA synthesis can be generally divided to heterogeneous and homogeneous processes. In the first process the HA formation involves adding a reagent of phosphorus such as phosphoric acid into a calcium hydroxide or other salt slurries. In the second method, calcium and phosphorus salts such as $Ca(NO_3)_2$ and $CaCl_2$ are dissolved in water under a controllable pH. Although, in both techniques a long time is needed to carry out the precipitation reactions, the appropriate reproducibility, high yield and simplicity make this synthesis method to be very competitive related to heterogonous processes.

Precipitation Technique

The most commonly used technique for the synthesis of HA is precipitation, involving wet chemical reactions between the calcium and phosphate precursors under controlled temperature and pH [24]. However, this method has some disadvantages such as difficult control of pH to avoid the formation of Ca-deficient HA as well as forming tricalcium phosphates (TCP) in calcination step. HA powder can be synthesized by precipitation method using orthophosphoric acid and calcium hydroxide [25]. The precipitated HA consists of small rod or platy particles. Aging the suspension leads to the growth of particles and changes the morphology. Calcium hydroxide, calcium phosphate, ammonium phosphate and phosphoric acid are normally used as reagents [26]. The secondary phase of CaO is observed on products synthesized by calcium hydroxide and calcium phosphate. HA and TCP are formed in the powder produced by ammonium phosphate. The crystallinity of the material is largely increased by thermal treatment and directly influences the formation of CaO and TCP phases. The amorphous phase remains in lower amount after thermal treatment [24, 26] as shown in Figure 1. The pure stoichiometric and thermally stable HA are easily obtained from the system $Ca(OH)_2$-H_3PO_4-H_2O at pH of 10.6, temperature of 95 °C and an appropriate calcination time [27]. The particles of obtained HA are tiny crystals with prism form. In synthesis of HA by the neutralization method, temperature is the most significant parameter affecting HA structural properties [28]. Room temperature and no aging of HA give a material with small crystallites and high specific surface area. The primary particle size decreases with higher agitation rates [29]. Different particle morphologies can be obtained at different agitation rates, depending on equipment selection.

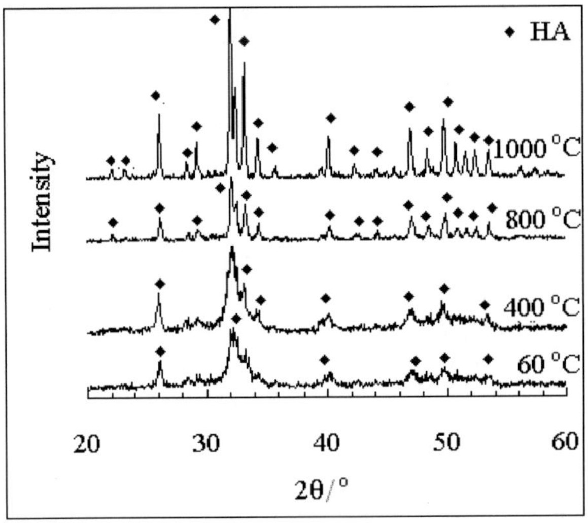

Figure 1. XRD patterns of HA powders calcined at different temperatures [24].

Nanoparticles of HA can be produced from $Ca(NO_3)_2{\cdot}4H_2O$ and $(NH_4)_2HPO_4$ by homogeneous precipitation [30]. The size of HA grains gradually increases when the sample is heated from 100 to 1200°C, and the hexagonaldipyramidal phase is not transformed to the other calcium phosphates phases up to 1200°C. Ca-deficient HA fibres can be synthesized by partly modifying HA processing method by a homogeneous precipitation technique. In the stoichiometric condition, the resulting fibres have long-axes of about 60–100 μm [31].

Sol-Gel Technique

The sol–gel technique is an appropriate method for preparation of pure powder due to the possibility of a strict control on process factors. This method was based on an ionic mixing of calcium and phosphorus to improve the chemical homogeneity. The sol–gel product is characterized by nano-size dimensions of primary particles. Moreover, the high reactivity of obtained powder allows a reduction of processing temperature and degradation phenomena occurring during the calcination. The major limitation related to sol–gel technique is the hydrolysis of phosphates and high cost of raw materials. HA nano-powder may be synthesized by the reaction of $Ca(NO_3)_2{\cdot}4H_2O$ and P_2O_5 using a simple sol–gel technique combined with heat-treatment at 900°C [32]. The obtained HA powder consists of nanoparticles in the range of 50–150 nm. HA powder can be prepared from calcium nitrate and ammonium hydrogen phosphate, adding etilendiamminetetraacetic acid (EDTA) and urea as gelling and ammonium agents [33]. The calcium ions are chelated by EDTA to permit a very good mixture with the phosphate ions and to prevent an immediate precipitate formation. The urea decomposition occurs in a very homogeneous system. The pure HA powder is finally achieved after a proper thermal treatment. In the sol–gel synthesis of HA, alkoxides or metal salts are frequently used as either calcium or phosphorus precursors. In most cases, phosphorus alkoxides are frequently employed as one of the major constituents, together with stoichiometric amount of calcium alkoxides or calcium salts, to form HA. The higher aging temperature and shorter time is required to form pure HA after calcination at 400-500°C. Impurity phases, such as CaO, $Ca_2P_2O_7$, $Ca_3(PO_4)_2$, $CaCO_3$, evolve if the aging parameters are not properly controlled [34].

Hydrothermal Technique

The conventional hydrothermal method has been applied for synthesis of HA but the control on morphology is poor. The addition of isopropanol to water as solvent influences the chemical composition and crystallinity of HA particles. The carbonate amount increases with increasing of isopropanol concentration [35]. In the presence of surfactants, the role of crystal surface is more important factor to control the nucleation. HA nanoparticles with uniform morphologies and controllable size can be synthesized by low temperature hydrothermal method in the presence of cationic surfactant. Cetyltrimethylammonium bromide (CTAB) is used to regulate the nucleation and crystal growth [36]. CTAB can bind with phosphate anion by the charge and stereochemistry complementarity, so that phosphate anion can be incorporated to the existing nuclei at a steady rate as a result, the final shape and size of HA particles can be well controlled. The temperature of reaction is more significant variable in altering the HA morphology and size. The nanoporous HA with pore size ranging from 1–5 nm can be produced by employment of CTAB. The pore structure of HA is thermally stable at temperature greater than 700°C and the pore size is independent of processing temperature and CTAB/PO_4 ratio [37].

Autoignition Technique

Recently, the autoignition technique has been paid increased interest to obtain nano-structured HA due to the easy operational conditions [38, 39]. The combustion synthesis is a wet chemical method, which generally is performed through the preparation of gel. In a typical process an aqueous solution containing the desired nitrate and phosphate ions with a suitable fuel is heated to form a complex gel. During the heating step, the gel undergoes self-combustion and an exothermic self-propagating reaction occurs, leading to the formation of a fluffy material. The autoignition method, which is based on the principles of propellant chemistry, a thermally induced redox reaction takes place between an oxidant and a fuel. The very weekly agglomerated HA powders can be synthesized by modified aqueous combustion reaction using urea and glycine as fuels [38]. The fuel type and fuel to oxidizer ratio significantly influence the characteristics of final product. The aqueous and alcohol media are employed in the presence of various fuels such as urea, glycine, alanine, hydrazine and hexamine. All the fuels used under the alcohol mediated combustion, result in pure HA formation [39].

Micelle Technique

The morphology of particles produced by precipitation, hydrothermal synthesis, sol–gel, and autoignition methods, are needle-like, sheet-like, or spherical shapes. The type of surfactant has great effect on the morphology of synthesized particles. HA nanopowders can be achieved using the micelle technique. Nanocrystalline HA is synthesized using dodecyl phosphate micelle system in which the surfactant concentration has an influence on nanoparticle surface area as well as morphology [40]. It was demonstrated that the synthesis of pure nanocrystalline HA using the dodecyl phosphate ($(C_{12}H_{25}O)P(O)(OH)_2$) micelle is between 60 and 80 nm with platelet morphology. The mixtures of cationic/anionic surfactants affect the morphology of HA nanoparticles due to electrostatic attraction between the oppositely charged head groups and hydrophobic interaction between the alkyl tails. The mixture of cetyltrimethyl ammonium bromide (CTAB) and sodium dodecyl sulfate (SDS) can be employed to obtain nanoparticles [41]. The morphology of the obtained powders at anionic-rich region (SDS:CTAB, 99:1) solution is rod like, but in the presence of cationic rich region (SDS:CTAB, 1:99) the resulted particles is sheet like which due to interaction of surfactants with opposite charges. The nano particles achieved in the presence of SDS are rod like but the morphology is less oriented than anionic-rich region.

Ultrasonic Technique

The HA nanoparticles can be prepared in an ultrasonic tubular microreactor [42]. The particle size and the aggregation degree are affected by the liquid flow rate in single-phase flow. The injection of gas reduces particle aggregation and decreases the size of primary particles, especially in higher liquid flow rates. Primary particles are smaller, particle shape is more homogeneous, crystallinity is higher, and the aggregation degree of the particles is reduced as well as carbonate contamination. The HA nano particles with a spherical morphology in the size of 30 nm can be synthesized using a hydrothermal chemical precipitation method incorporating low power sonic irradiation [43]. The crystallinity and morphology of HA depends on Ca/P ratio, ultrasonic irradiation and temperature.

Microwave Technique

Microwave synthesis causes the efficient transformation of energy in a shorter period. The HA nanoparticles can be synthesized by microwave treatment of calcium chloride ($CaCl_2$) and sodium phosphate (Na_3PO_4) in solid phase [44].

Environmental Applications of Nano-Sized Hydroxyapatite

Removal of Heavy Metal Cations

The removal of Cd^{2+} and Zn^{2+} from aqueous solutions can be carried out by Fe-HA nanoparticles [45]. The adsorption kinetic data are well fitted by the pseudo-second order model and adsorption isotherms are best described by the Langmuir model [46]. The adsorption amount increases with increase in pH from 4 to 8. In the binary component system, the total adsorption capacity is approximately constant, but the nano-sized HA adsorbent have better affinity for cadmium ion than for zinc ion. The higher achieved uptake is particularly attractive for the practical use of HA in the decontamination of solutions polluted by cadmium [47]. The HA synthesized by homogenous precipitation method can be employed for the removal of cadmium ions from aqueous solutions. In the Cd-Pb-Zn-Cu system, the competitive adsorption among the heavy metals reduces the amount of adsorbed Cd compared to the single-metal system [48]. The equilibrium uptake is increased with increasing the initial concentration of metal ions in solution [49]. The ion adsorption capacity of HA increases with decreasing crystallinity and increasing specific surface area [50]. A further increase of the calcination temperature to 1000°C results in a strong decrease in the lead removal capacity. As shown in Figure 2, since the surface area of nanocrystalline HA is significantly high, this adsorbent can be of interest to effectively immobilise heavy metal ions from wastewater. The cadmium removal by HA is mainly controlled by sample specific surface area, which is directly linked to the crystallite dimensions [51]. The Cd^{2+} uptake by noncalcined and calcined HA samples follows both Langmuir isotherms, reinforcing the hypothesis that surface adsorption sites control the adsorption process.

HA is capable for removing Al^{3+}, Co^{2+}, Cr^{3+}, Cu^{2+}, Fe^{3+}, Ni^{2+} and Pb^{2+} ions from solution [52]. For solutions containing a single ionic species, the selectivity appears to be $Cr^{3+} > Pb^{2+} > Fe^{3+} > Al^{3+} > Co^{2+} > Cu^{2+} > Ni^{2+}$. When solution contains three types of ions, the electronegativity plays an important role in uptake of ion. If the ions indicate the same electronegativity, the valence of ions determines the adsorption efficiency. It is believed that the increase in impurity of HA powder increases the number of lattice defects which are ideal for adsorption. The powders with high specific surface area are superior to use as an adsorbent as a result, a low sintering temperature is preferable. A low pH encourages initial dissolution of HA which increases the removal capacity. With increasing process temperature the diffusion rate of the ions increases, leading to higher uptake of ions.

(a) (b)

Figure 2. SEM micrographs of HA powders (a) noncrystalline and (b) calcined at 1000°C [24].

Binary-metal system induces competitive adsorption among the metals with reduction of removal capacity compared to the single metal adsorption [53]. The Cu and Zn immobilization mechanism was through a two-step process: the first step is the surface complexation on specific sites of HA surface. The second step is an ion exchange with Ca^{2+} and the precipitation of a heavy metal containing HA. The reduced efficiency of HA in binary systems should be carefully considered for practical environmental applications, processing wastewater containing heavy metals.

The variations in NO_3^- and Cl^- concentrations have no effect on Pb^{2+} immobilization by HA, whereas F^- and CO_3^{2-} decrease Pb^{2+} immobilization [54]. There is internal competition between the heavy metal ions for precipitation and adsorption in active sites [55]. The Pb removal mechanism from solution by HA remains unchanged in the presence of aqueous Cd and Zn. However, the competitive effects among the heavy metals may alter the adsorption behaviors of Cd and Zn.

HA is effective in immobilizing lead under an insoluble form (pyromorphite), following ion exchange between apatite Ca^{2+} and Pb^{2+} ions in solution [56]. Although partial dissociation of HA leads to the formation of soluble lead phosphates in acidic pH however these compounds are present in negligible amounts when the weight ratio between HA and lead is high. The Pb^{2+} adsorption mechanism is different respect to the mechanism of Cd^{2+}, Zn^{2+}, and Sr^{2+} adsorption. The removal of Pb^{2+} from the solution is almost complete at pH range of 3–12. The adsorption of other cations generally increases with the rise in pH. Uptake of Pb^{2+} is rapid in comparison to other cations. Metal immobilization is followed by final pH decrease and linear increase of Ca^{2+} concentration. HA selectively adsorbs metal ions from the solutions commonly present in the environment like Ca^{2+}, Mg^{2+}, Na^+ and K^+. Neither of them influences Pb^{2+} immobilization, while presence of Ca^{2+} slightly reduces adsorption of other cations in the order $Sr^{2+} > Cd^{2+} > Zn^{2+}$.

Removal of Dyes

The poorly crystalline HA can remove Disperse Blue SBL dye from aqueous solutions [57]. The initial rate of adsorption is high. The adsorption is greatly pH dependent, with a high uptake of dye at low pH. The poorly crystalline shows greater potential for the removal of textile disperse dyes. Carbonate substituted nano HA is found to be an efficient and low cost adsorbent for the removal of reactive dyes from aqueous solution.

Uncalcined HA can be employed as adsorbent for removing Acid Black 172 dye from aqueous solution [58]. The dye adsorption follows a pseudo-second order kinetic model with intraparticle diffusion as one of the rate determining steps. The Langmuir isotherm model provides the best fit of the equilibrium data. The adsorption process is spontaneous and decreases with increase in temperature, showing the exothermic nature of adsorption. The Acid Black 172 dye adsorption on HA may be explained to proceed via the electrostatic attraction between the positively charged surface of HA and the negatively charged groups of dye. The calcined nano HA produced by homogenous precipitation method indicates the good adsorption ability towards crystal violet dye and can be used as a low cost adsorbent for removing the dyes from wastewater.

Manufacturing Adsorbent Bed

The random packed columns filled with random shaped HA are widely used for wastewater treatment. Many processing routes including, slip casting, pressing, extrusion and granulation can be considered as devices for forming random HA packed beds. Extrusion is one of the shaping techniques which is the most suitable method for producing a variety of shapes with constant cross section such as Raschig ring. The advantages of this method are the suitable strength of specimens after forming, accurate control of dimensions in the desired level and rapid manufacturing technique. This process can be upgraded to industrial scale to get suitable rings. Regardless the adsorption capacity, these characteristics largely determine the mechanical properties and performance of packed bed [59].

The mechanical reliability of Raschig rings is one of the important key parameters for the effective performance of packed bed. The low compressive strength may be cause to break the adsorbents during fabricating process or application in the industrial scale. This phenomenon makes various operational problems for industrial units such as blockages and increase pressure drop. Before the employment of adsorbent in contaminant treatments, the formed HA should be calcined [60, 61].

The new adsorbents can be produced by mixing uncalcined HA and cement kiln dust (CKD) shaped by an extruder to use in adsorption of lead from aqueous solutions. As illustrated in Figure 3, the addition of 50 wt.% HA into the adsorbent composition cannot effectively change the adsorption capacity [62]. The presence of HA in adsorbent composition increases the strength two times in comparison to the bed prepared with CKD only. Also, the mechanical reliability of adsorbent is not affected by CKD.

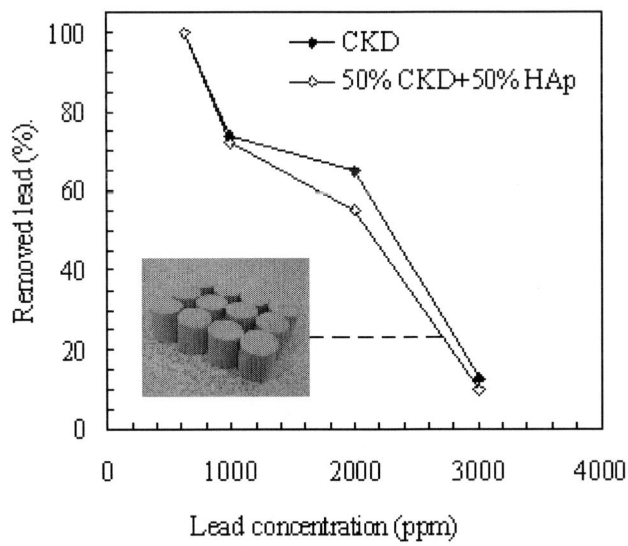

Figure 3. Removal of lead by nano-sized HA and CKD [62].

CONCLUSION

This chapter attempted to cover a wide range of HA synthesis techniques. The reader can get an idea about the various types of HA used for the removal of heavy metal ions and organic dyes from the wastewater. The study was carried out in the direction to understand the mechanism of adsorption. The kinetic data are better described by the pseudo-second-order model and the adsorption isotherm can be well fitted by Langmuir equation. However, the synthesis method, calcination condition and the presence second cation in HA structure affects the heavy metal ion and dye uptake. Thus uptake efficiency is accompanied by mechanism of ion exchange as well as adsorption. Most of the reported studies are performed in the batch process and the further research is to be carried out to make the process economically viable in the industrial scale with focus on heavy metal ions and organic dyes. The future researches might be accompanied by adsorption-desorption process to fulfill the goal of zero waste.

REFERENCES

[1] T. A. Kurniawan, G. Y. S. Chan, W. H. Lo and S. Babe, "Physico-chemical treatment techniques for wastewater laden with heavy metals," *Chem. Eng. J.*, vol. 118, no. 1-2, pp. 83–98, 2006.

[2] A. Mohammad and N. P. A., Mohamed, "Physico-chemical adsorption treatments for minimization of heavy metal contents in water and wastewaters," *J. Sci. Ind. Res.*, vol. 56, pp. 523–539, 1997.

[3] A. Gurses, M. Yalcin and C. Dogar, "Electrocoagulation of some reactive dyes: a statistical investigation of some electrochemical variables," Waste Manage., vol. 22, no. 5, pp. 491-499, 2002.

[4] O. J. Hao, H. Kim and P. C. Chiang, "Decolourization of wastewater," *Crit. Rev. Environ. Sci. Technol.*, vol. 30, pp. 449-505, 2000.

[5] Y. M. Slokar and A. M. L. Marechal, "Methods of decoloration of textile wastewaters," *Dyes Pigments*, vol. 37, pp. 335-356, 1998.

[6] S. J. T. Pollard, G. D. Fowler, C. J. Sollars and R. Perry, "Low-cost adsorbents for waste and wastewater treatment: a review," *Sci. Total Environ.*, vol. 116, no. 1-2, pp. 31-52, 1992.

[7] A. Salem, H. Afshin and H. Behsaz, "Removal of lead by using Raschig rings manufactured with mixture of cement kiln dust, zeolite and bentonite," *J. Hazard. Mater.*, vol. 223– 224, pp. 13– 23, 2012.

[8] A. Salem and R. Akbari Sene, "Removal of lead from solution by combination of natural zeolite-kaolin- bentonite as a new low-cost adsorbent," *Chem. Eng. J.*, vol. 174, pp. 619– 628, 2011.

[9] A. Salem and R. Akbari Sene, "Optimization of zeolite-based adsorbent composition for fabricating reliable Raschig ring shaped by extrusion using Weibull statistical theory," *Micropor. Mesopor. Mater.*, vol. 163, pp. 65–75, 2012.

[10] A. Salem and L. Karimi, "Physico-chemical variation in bentonite by sulfuric acid activation," *Korean J. Chem. Eng.*, vol. 26, no. 4, pp. 980-984, 2009.

[11] L. Karimi and A. Salem, "The role of bentonite particle size distribution on kinetic of cation exchange capacity," *J. Ind. Eng. Chem.*, vol. 17, pp. 90–95, 2011.

[12] A. Salem, M. Saghapour, "Relationship between acidification factors of Ca-bentonite and methylene blue uptake: optimization and kinetic study," *Chem. Papers*, DOI: 10.1515/chempap-2015-0195, 2015.

[13] Sh. Salem, A. Salem and A. Agha Babaei, "Preparation and characterization of nano porous bentonite for regeneration of semi-treated waste engine oil: Applied aspects for enhanced recovery," *Chem. Eng. J.*, vol. 260, pp. 368–376, 2015.

[14] Sh. Salem, A. Salem and A. Agha Babaei, "Application of Iranian nano-porous Ca-bentonite for recovery of waste lubricant oil by distillation and adsorption techniques," *J. Ind. Eng. Chem.*, vol. 23, pp. 154–162, 2015.

[15] I. Smiciklas, A. Onjia, S. Raicevic, D. Janackovic and M. Mitric, "Factors influencing the removal of divalent cations by hydroxyapatite," *J. Hazard. Mater.*, vol. 152, pp. 876–884, 2008.

[16] G. Ciobanu, M. Harja, M. Diaconu, C. Cimpeanu, R. Teodorescu and D. Bucur, "Crystal violet dye removal from aqueous solution by nanohydroxyapatite," *J. Food, Agriculture Environ.*, vol. 12, no. 1, pp. 499-505, 2014.

[17] S. Kannan, J. H. G. Rocha, J. M. G. Ventura, A. F. Lemos and J. M. F. Ferreira, "Effect of Ca/P ratio of precursors on the formation of different calcium apatitic ceramics—An X-ray diffraction study," *Scripta Mater.*, vol. 53, pp. 1259–1262, 2005.

[18] R. I. Martin and P.W. Brown, "Aqueous formation of hydroxyapatite," *J. Biomed. Mater. Res.*, vol. 35, pp. 299–308, 1997.

[19] T. S. B. Narasaraju and D. E. Phebe, "Review some physico-chemical aspects of hydroxylapatite," *J. Mater. Sci.*, vol. 31, pp. 1-21, 1996.

[20] A. Osaka, Y. Miura, K. Takeuchi, M. Asada and K. Takahashi, "Calcium apatite prepared from calcium hydroxide and orthophosphoric acid," *J. Mater. Sci.: Mater. Med.*, vol. 2, pp. 51–55, 1991.

[21] I. R. Gibson, I. Rehman, S. M. Best and W. Bonfield, "Characterization of the transformation from calcium-defcient apatite to b-tricalciumphosphate," *J. Mater. Sci.: Mater. Med.*, vol. 12, pp. 799-804, 2000.

[22] K. Ishikawa, P. Ducheyne and S. Radin, "Determination of the Ca/P ratio in calciumdeficient hydroxyapatite using X-ray diffraction analysis," *J. Mater. Sci.: Mater. Med.*, vol. 4, pp. 105-168, 1993.

[23] R. K. Kramer, A. M. Morey, M. Staruch, M. S. Suib, M. Jain, J. I. Budnick and M. Wei, "Synthesis and characterization of iron-substituted hydroxyapatite via a simple ion-exchange procedure," *J. Mater. Sci.*, vol. 48, pp. 665–673, 2013.

[24] A. Salem and E. Velayi, "Factors influencing the thermal stability and lead sorption capacity of hydroxyapatite manufactured by precipitation method using industrial calcium hydroxide," *Int. J. Appl. Ceram. Technol.*, vol. 10, no. 4, pp. 643–654, 2013.

[25] M. R. Saeri, A. Afshar, M. Ghorbani, N. Ehsani and C. C. Sorrell, "The wet precipitation process of hydroxyapatite," *Mater. Lett.*, vol. 57, pp. 4064–4069, 2003.

[26] M. H. Santos, M. Oliveira, L. P. F. Souza, H. S. Mansur and W. L. Vasconcelos, "Synthesis control and characterization of hydroxyapatite prepared by wet precipitation process," *Mater. Res.*, vol. 7, no. 4, pp. 625-630, 2004.

[27] S. Lazid, J. Katanid-Popovid, S. Zec and N. Miljevid, "Properties of hydroxyapatite crystallized from high temperature alkaline solutions," *J. Crystal Growth,* vol. 165, pp. 124-128, 1996.
[28] I. Smiciklas, A. Onjia and S. Raicevic, "Experimental design approach in the synthesis of hydroxyapatite by neutralization method," *Separation and Purification Technol.*, vol. 44, pp. 97–102, 2005.
[29] M. N. Salimi, R. H. Bridson, L. M. Grover and G. A. Leeke, "Effect of processing conditions on the formation of hydroxyapatite nanoparticles," *Powder Technol.*, vol. 218, pp. 109–118, 2012.
[30] I. Mobasherpour, M. Soulati Heshajin, A. Kazemzadeh and M. Zakeri, "Synthesis of nanocrystalline hydroxyapatite by using precipitation method," *J. Alloys Compd.*, vol. 430, pp. 330-333, 2007.
[31] M. Aizawa, H. Ueno, K. Itatani and I. Okada, "Syntheses of calcium-deficient apatite fibres by a homogeneous precipitation method and their characterizations," *J. Eur. Ceram. Soc.*, vol. 26, pp. 501–507, 2006.
[32] I. S. Kim and P. N. Kumta, "Sol–gel synthesis and characterization of nanostructured hydroxyapatite powder," *Mater. Sci. Eng. B*, vol. 111, pp. 232–236, 2004.
[33] G. Bezzi, G. Celotti, E. Landi, T. M. G. Torretta, I. Sopyan and A. Tampieri, "A novel sol–gel technique for hydroxyapatite preparation," *Mater. Chem. Phys.*, vol. 78, pp. 816–824, 2003.
[34] D. M. Liu, T. Troczynski and W. J. Tseng, "Aging effect on the phase evolution of water-based sol–gel hydroxyapatite," *Biomater.*, vol. 23, pp. 1227–1236, 2002.
[35] R. N. Panda, M. F. Hsieh, R. J. Chung and T. S. Chin, "FTIR, XRD, SEM and solid state NMR investigations of carbonate-containing hydroxyapatite nano-particles synthesized by hydroxide-gel technique," *J. Phys. Chem. Solids*, vol. 64, pp. 193–199, 2003.
[36] Y. Wang, S. Zhang, K. Wei, N. Zhao, J. Chen and X. Wang, "Hydrothermal synthesis of hydroxyapatite nanopowders using cationic surfactant as a template," *Mater. Lett.*, vol. 60, pp. 1484–1487, 2006.
[37] Y. Li, W. Tjandra and K. C. Tam, "Synthesis and characterization of nanoporous hydroxyapatite using cationic surfactants as templates," *Mater. Res. Bull.*, vol. 43, pp. 2318–2326, 2008.
[38] S. K. Ghosh, S. K. Roy, B. Kundu, S. Datta and D. Basu, "Synthesis of nano-size hydroxyapatite powders through solution combustion route under different reaction conditions," *Mater. Sci. Eng. B*, vol. 176, pp. 14–21, 2011.
[39] R. Ramakrishnan, P. Wilson, T. Sivakumar and I. Jemina, "A comparative study of hydroxyapatites synthesized using various fuels through aqueous and alcohol mediated combustion routes," *Ceram. Int.*, vol. 39, pp. 3519–3532, 2013.
[40] Y. Wu and S. Bose, "Nanocrystalline hydroxyapatite: micelle template synthesis and characterization," *Langmuir,* vol. 21, pp. 3232-3234, 2005.
[41] N. Esmaeilian Tari, M. M. Kashani Motlagh and B. Sohrabi, "Synthesis of hydroxyapatite particles in catanionic mixed surfactants template," *Mater. Chem. Phys.*, vol. 131, pp. 132–135, 2011.
[42] F. Castro, S. Kuhn, K. Jensen, A. Ferreira, R. Rocha, A. Vicente and J. A. Teixeira, "Process intensification and optimization for hydroxyapatite nanoparticles production," *Chem. Eng. Sci.*, vol. 100, pp. 352–359, 2013.

[43] G. E. Poinern, R. K. Brundavanam, N. Mondinos and Z. T. Jiang, “Synthesis and characterisation of nanohydroxyapatite using an ultrasound assisted method,” *Ultrasonics Sonochemistry*, vol. 16, pp. 469-474, 2009.

[44] P. Parhi, A. Ramanan and A. R. Ray, “A convenient route for the synthesis of hydroxyapatite through a novel microwave-mediated metathesis reaction,” *Mater. Lett.*, vol. 58, no. 27–28, pp. 3610–3612, 2004.

[45] Y. Feng, J. L. Gong, G. M. Guang-Ming Zeng, Q. Y. Niu, H. Y. Zhang, C. G., Niu, J. H. Deng and M. Yan, “Adsorption of Cd (II) and Zn (II) from aqueous solutions using magnetic hydroxyapatite nanoparticles as adsorbents,” *Chem. Eng. J.*, vol. 162, pp. 487–494, 2010.

[46] R. Zhu, R. Yu, J. Yao, D. Mao, C. Xing and D. Wang, “Removal of Cd^{2+} from aqueous solutions by hydroxyapatite,” *Cata. Today*, vol. 139, pp. 94–99, 2008.

[47] S. McGrellis, J. N. Serafini, J. JeanJean, J. L. Pastol and M. Fedoroff, “Influence of the sorption protocol on the uptake of cadmium ions in calcium hydroxyapatite,” *Sep. Purif. Technol.*, vol. 24, pp. 129–138, 2001.

[48] A. Corami, S. Mignardi and V. Ferrini, “Cadmium removal from single- and multi-metal (Cd+Pb+Zn+Cu) solutions by sorption on hydroxyapatite,” *J. Colloid Interface Sci.*, vol. 317, pp. 402-408, 2008.

[49] N. Noureddine, Q. Samir, A. Ali, N. Abederrahman, A. I. Yhya, “Kinetics and equilibrium of cadmium removal from aqueous solutions by sorption onto synthesized hydroxyapatite,” *Desalin. Water. Treat.*, vol. 43, pp. 8–16, 2012.

[50] C. Stotzel, F. A. Muller, F. Reinert, F. Niederdraenk, J. E. Barralet and U. Gbureck, “Ion adsorption behaviour of hydroxyapatite with different crystallinities,” *Colloids Surf. B: Biointerfaces*, vol. 74, pp. 91–95, 2009.

[51] N. C. C. Darocha, R. Decampos, A. M. Rossi, E. L. Moreira and A. T. Barbosa, “Cadmium uptake by hydroxyapatite synthesized in different conditions and submitted to thermal treatment,” *Environ. Sci. Technol.*, vol. 36, pp. 1630-1635, 2002.

[52] J. Reichert and J. G. P. Binner, “An evaluation of hydroxyapatite-based filters for removal of heavy metal ions from aqueous solutions,” *J. Mater. Sci.*, vol. 31, pp. 1231-1241, 1996.

[53] A. Corami, S. Mignardi and V. Ferrini, 2007. “Copper and zinc decontamination from single- and binary-metal solutions using hydroxyapatite,” *J. Hazard. Mater.*, vol. 146, pp. 164–170, 2007.

[54] Q. Y. Ma, T. J. Logan and S. J. Traina, “Effects of NO_3^-, Cl^-, F^- SO_4^{2-}, and CO_3^{2-} on Pb^{2+} immobilization by hydroxyapatite,” *Environ. Sci. Technol.*, vol. 28, pp. 408-418, 1994.

[55] X. Chen, J. Wright, J. Conca and L. Peurrung, “Effects of pH on heavy metal sorption on mineral apatite,” *Environ. Sci. Technol.*, vol. 31, no. 3, pp. 624-631, 1997.

[56] N. Arnich, M. C. Lanhers, F. Laurensot, R. Podor, A. Montiel and D. Burnel, “In vitro and in vivo studies of lead immobilization by synthetic hydroxyapatite,” *Environ. Pollution*, vol. 124, pp. 139–149, 2003.

[57] B. Noureddine, Q. Samir, A. Ali, N. Abederrahman and A. I. Yhya, Adsorption of Disperse Blue SBL dye by synthesized poorly crystalline hydroxyapatite, *J. Environ. Sci.*, vol. 20, pp. 1268–1272, 2008.

[58] G. Ciobanu, M. Harja, L. Rusu, A. M. Mocanu and C. Luca, "Acid Black 172 dye adsorption from aqueous solution by hydroxyapatite as low-cost adsorbent," *Korean J. Chem. Eng.*, vol. 31, no. 6, pp. 1021-1027, 2014.

[59] J. M. Villora, P. Callejas, M. F. Barba and C. Baudin, "Statistical analysis of the fracture behaviour of porous ceramic Raschig rings. *J. Eur. Ceram. Soc.*, vol. 24, pp. 589–594, 2004.

[60] S. Bailliez and A. Nzihou, "The kinetics of surface area reduction during isothermal sintering of hydroxyapatite adsorbent," *Chem. Eng. J.*, vol. 98, pp. 141–152, 2004.

[61] M. G. S. Murray, J. Urray, J. Wang, C. B. Ponon, P. M. Marquis, "An improvement in processing of hydroxyapatite ceramics," *J. Mater. Sci.*, vol. 30, pp. 3061-3074, 1995.

[62] A. Salem and E. Velayi, "Application of hydroxyapatite and cement kiln dust mixture in adsorption of lead ions from aqueous solution," *J. Ind. Eng. Chem.*, vol. 18, pp. 1216–1222, 2012.

In: High Value Processing Technologies
Editors: A. Berenjian, H. Jafarizadeh-Malmiri et al.
ISBN: 978-1-63484-453-6

Chapter 6

THE EFFECTIVENESS OF MICROBIAL CRACK TREATMENT IN SELF HEALING CONCRETE

***Mostafa Seifan*[1], *Ali Khajeh Samani*[1], *Jeffrey John Burgess*[2] *and Aydin Berenjian*[1,*]**
[1]School of Engineering, Faculty of Science and Engineering,
The University of Waikato, Hamilton, New Zealand
[2]Supacrete Concrete Ltd, Tauranga, New Zealand

ABSTRACT

Concrete is material of choice for many applications due to its availability, high compressive strength and relatively low cost. Despite its advantages, it is susceptible to cracking under internal and external stresses. Penetration of aggressive substances through cracks has a detrimental impact on concrete service life by corrosion of reinforcement bars. Different types of crack treatments such as intrinsic, external coating and internal healing are currently under research. In recent years, calcium carbonate precipitation through embedded microbial sources in the concrete matrix has drawn many attention. It is concluded that biologically healing approach can substantially enhance durability of concrete. This chapter is aimed to evaluate the effectiveness of bio self-healing approach and also prospective challenges in microbial crack treatment are discussed.

INTRODUCTION

Concrete is one of the most used materials in the world; every year billions of tonnes are produced and consumed. Unique properties, such as being durable, recoverable, efficient, cost-effective and having a high compressive strength capacity, make concrete one of the most used materials in the construction industry. Concrete is widely used in highway bridges,

* Corresponding author: Dr. Aydin Berenjian, School of Engineering, Faculty of Science and Engineering, The University of Waikato, Hamilton, New Zealand, Email: aydin.berenjian@waikato.ac.nz.

underground structures, jetties, tunnels, dams and buildings. It can be easily moulded into a desired size and shape or pumped to construct high structures. Concrete recycling is used in many countries to reduce waste, landfill degradation and transportation costs as well as reduce the need for mining. The existing concrete structure is crushed into small particle sizes and reused as an aggregate in ready mixed concrete, soil stabilisation and road foundation.

Concrete is primarily composed of cement, water and aggregate (fine and coarse). Cement characteristics depend on the relative amount of its composition. Typical composition limits for Portland cement can be defined from Table 1, which consists of additional minor compounds [1]. To enhance its mechanical properties and modify characteristics, admixtures, such as plasticisers, superplasticisers, accelerators, set-retarders, water-reducing and air entraining admixtures, often mix with other compositions. Aggregates bind together as the cement reacts with water, which is known hydration. There are different types of cement which are broadly divided into two groups: hydraulic and non-hydraulic. Hydraulic cement requires water to hydration, whereas carbon dioxide causes non-hydraulic cement to solidify [1]. Based on required purposes, different types of cements, such as ordinary, modified, rapid-hardening, low-heat, sulphate resisting, high slag blast-furnace, white and pozzolana Portland, are produced. Raw materials (limestone, silica, alumina and iron ore) are transported from quarry to cement plants. In the plant the materials need to be grinded into a fine powder and heated to temperature of 1400°C, causing the mixture to break down and form a new compound.

Cement production results in significant environmental and sustainability issues. Cement industries attribute to one-tenth of carbon dioxide emissions as the main cause of increasing greenhouse gas emissions [2]. It has increased from 62.4 million tonnes in the 1900s to 4.18 billion tonnes in 2014 [3, 4]. China, as one of the biggest concrete consumers in the world, consumed more than 6.6 gigatonnes of cement between 2011 and 2013 [5]. As shown in Figure 1, the current condition has a detrimental effect on air quality throughout cement manufacturing and the delivery process.

Strength and durability are two significant characteristics which have direct impact on concrete service life. Although concrete has a high compressive capacity, it has low tensile strength, which is measured by flexural test. Concrete structures have experienced flexure or bending stresses during their life time. Therefore it is reinforced with steel bars, typically in the form of steel rods lain throughout the structure. Steel has high tensile strength and it is also compatible with concrete compositions. Ideally, proper reinforcement and concrete design causes to increase concrete lifespan up to 70 years.

Table 1. Approximate composition limits of Portland cement (weight %) [1]

Oxide	Content percent
CaO	60-67
SiO_2	17-25
Al_2O_3	3-8
Fe_2O_3	0.5-6.0
MgO	0.1-4.0
Alkalis	0.2-1.3
SO_3	1-3

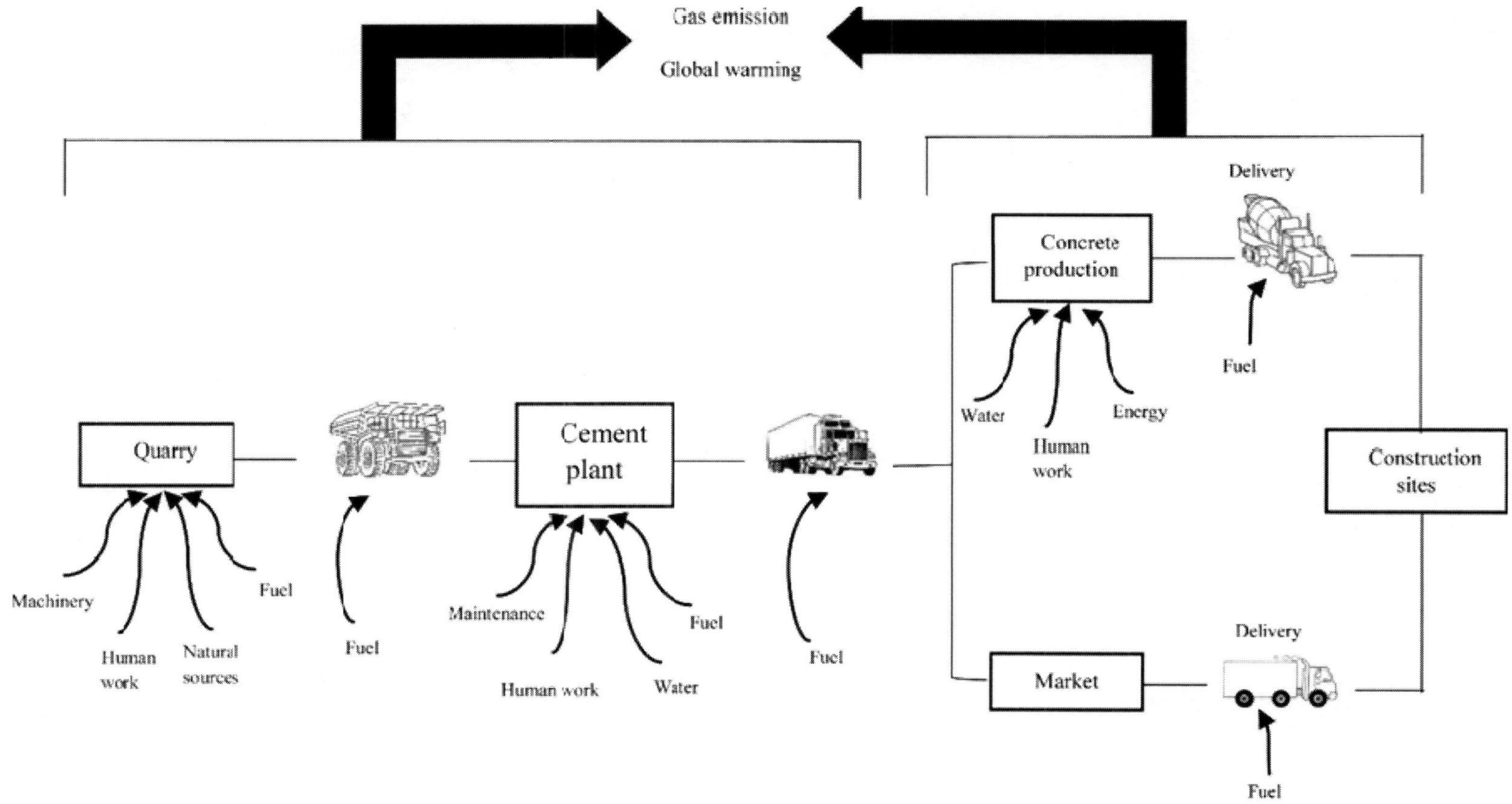

Figure 1. Energy consumption and gas emission in cement production cycle.

Concrete durability is defined as the ability of concrete to resist water or chemical penetration. Cracks have adverse impact on durability of concrete and they can be formed in both plastic and hardened stages. There are two types of cracks in respect to their width, namely structural and non-structural cracks (surface cracks). Non-structural and surface cracks are less likely to cause structural failure; therefore they can be repaired by applying sealants or grouts. Whereas those cracks which have the potential of becoming destructive are mostly repaired by stitching, doweling, and in some cases replacement of the structure [6].

Various conditions, such as expansion and contact upon wetting and drying cycles, shrinkage on drying fresh concrete, external loading, temperature changes, poor construction and internal stresses (mostly caused by uneven settlement), result in the formation of cracks. Furthermore, in cold regions, frozen water trapped in a matrix causes increasing fluid pressure and therefore cracks will appear due to exceeded tensile strength [7].

By increasing the size of capillary pores and cracks, the flow rate of fluids or gasses into concrete is increased. Crack generation and consequently corrosion through reaction of destructive substances with concrete composition and reinforcement bars results in a significant reduction in concentre service life. Chloride ion attack as a major type of ionic diffusion has a destructive impact on reinforcement bars by accelerating corrosion. Since the bar erodes, the rust occupies more space within the concrete structure, exceeding the volume occupied by the reinforcement bar. This phenomenon results in tensile stress and consequently paves the path for spalling, crack formation and concrete delamination [8-11].

TREATMENT APPROACHES

To date, different techniques to repair concrete degradation have been introduced. The healing agent can be applied either externally (on the concrete surface) or internally (within concrete structure), which are known as passive treatment and active treatment, respectively. There are a number of important differences between active and passive treatments. Passive treatments are used for the cracks in exterior sides and those parts of structures which are reachable. Periodic inspection and manual insertion, as well as negative influence of the filler on the environment are the main shortcomings associated with passive treatments. Whereas active healing methods can operate throughout the structure regardless of the position of cracks.

There are different forms of concrete healing techniques that are currently under research: intrinsic, external coating and internal healing. Each method of healing differs in the mechanism used to repair the crack in the concrete.

Autogenous

Intrinsic healing in concrete, which is known as autogenous healing, occurs due to the composition of the concrete. It is reported that concrete has the ability of filling cracks by producing calcium carbonate through the hydration of un-hydrated cement particles or carbonation of dissolved calcium hydroxide [12]. Younger concrete tends to experience the hydration of un-hydrated cement particles that still exist in the concrete. As shown in Eq. 1, calcium hydroxide is produced with reaction of water and calcium oxide present in concrete. Then reaction of calcium hydroxide and atmospheric carbon dioxide results in calcium carbonate production (Eq. 2) [13].

$$CaO + H_2O \rightarrow Ca(OH)_2 \tag{1}$$

$$Ca(OH)_2 + CO_2 \rightarrow CaCO_3 + H_2O \tag{2}$$

It has been noted that autogenous healing tends to favour narrow cracks ranging from 0.1 to 0.3 mm [14-17]. Autogenous healing highly depends on the position and quality of un-hydrated cement, the age of the concrete and presence of water. Mostly, un-hydrated cement particles are located away from the surface of the concrete where most of cracking occurs. Although reduction in water to cement ratio (w/c) can enhance autogenous healing capacity, it has a negative impact on shrinkage and workability of concrete [2, 18]. The use of hydrogels is another way in which autogenous healing can be enhanced. The hydrogels are placed in the concrete during mixing. However, the empty pores that are created upon rupture of hydrogels weaken the structural integrity of the concrete.

Despite the fact that autogenous healing can fill the crack, the hydration of un-hydrated cement may cause cracks in large volume placement. Exothermic chemical reaction between un-hydrated cement and water may generate significant heal and raise the concrete temperature. If the temperature gradient between the outer part and inner part of the concrete structure is significant, the likelihood of cracking increases.

External Coating

In the second group, repairing compounds are applied on the concrete surface. Currently, chemical and polymer sealers are used due to their availability, flexibility to move into cracks and low price. Chemical compound is injected into the crack to restore the structural integrity of the concrete by filling the cracks. Flexibility of polymers as an alternative for chemical injection encourages consumers to use this high adhesive materials. Cracks are filled as the polymer is injected into the cracks and expands upon presence of moisture.

However, utilisation of sealers are associated with several drawbacks. This technique requires labour to detect cracks and examine the concrete integrity and also the application of some of the fillers, such as epoxy polyurethane, acrylics and siloxane which have an adverse impact on the environment. As the chemical treatments are not permanent, the repair should be repetitive, and that results in high maintenance cost. Poor weathering resistance, sensibility to temperature changes, low heat resistance, moist sensitivity, poor bonding, and susceptibility to degradation and delamination with age are attributed to the external coatings [19-21].

Internal Healing

Internal healing involves incorporation of a healing agent into a concrete matrix during concrete preparation. Encapsulation is a common practice to incorporate a healing agent in a concrete matrix. These capsules are tiny in size (micro-scale) and contain chemicals or polymers. Dry [22] performed pioneering method of self-healing by releasing chemicals from incorporated hollow fibres inside cementitious material. Released healing agent upon capsule rupture reacts with water or moisture, concrete compositions or secondary component embedded in concrete to fill the cracks.

This consequently leads to produce filler that plugs the crack, porosity and void. However, the capsules and remaining space after their activation weaken the concrete integrity. Furthermore, incorporated agent cannot be viable during concrete lifespan or even lost its activity under environmental conditions. Another pathway in which concrete can heal itself is by solid debris that comes in with the water as it seeps into the crack and it will plug up the crack. However, the water that did get into the concrete ahead of the solid debris will tend to damage the concrete by corrosion of metal reinforcement lain in the structure.

Due to the aforementioned shortcomings associated in current crack treatments, the need for research into controlled and sustainable healing techniques are in demand. In recent years bio self-healing technique has drawn much interest. Bio self-healing concrete can diagnose itself and react to any damage in the form of cracking that occurs. The bio self-healing technique helps to reduce maintenance and repair costs by means of producing a durable long lasting structure and it also helps to reduce carbon dioxide emission. In this method, calcium carbonate is produced through biomineralisation to fill the cracks.

BIOMINERALISATION OF CALCIUM CARBONATE

Calcium carbonate as the most abundant mineral comprises about 4% of the earth's crust as limestone form. Calcium carbonate is widely used as an additive and filler in different industries, such as food and construction, due to its physical and chemical characteristics [23, 24]. Calcium carbonate is one of the most suitable fillers in concrete due to high compatibility with cementitious compositions. Calcium carbonate is mostly extracted by mining. However, carbonation of calcium hydroxide (Eq. 3 and 4), synthesis of inorganic materials [25] (Eq. 5) and biomineralisation are another pathways to produce calcium carbonate.

$$CaO + H_2O \rightarrow Ca(OH)_2 \quad (3)$$

$$Ca(OH)_2 + CO_2 \rightarrow CaCO_3 + H_2O \quad (4)$$

$$Ca^{2+} + MgCO_3 \rightarrow CaCO_3 + Mg^{2+} \quad (5)$$

Biomineralisation is the process which minerals are formed by living organisms. Microbial activity plays a key role in nature, through nutrient cycling and biogeochemical cycling of nitrogen, sulphur, iron, carbon and manganese [26]. Biomineralisation of calcium carbonate can be established through (i) biologically induced mineralisation (BIM) and (ii) biologically controlled mineralisation (BCM). Biologically controlled mineralisation results in well-defined compositions and structures minerals. Most of biologically controlled mineralisation processes occur in intracellular sites, extracellular sites or a combination of them to achieve a higher order of assembly [27]. Whereas BIM usually occurs in an open environment and formation of minerals relies on reaction of metabolic products and surrounding environments [28].

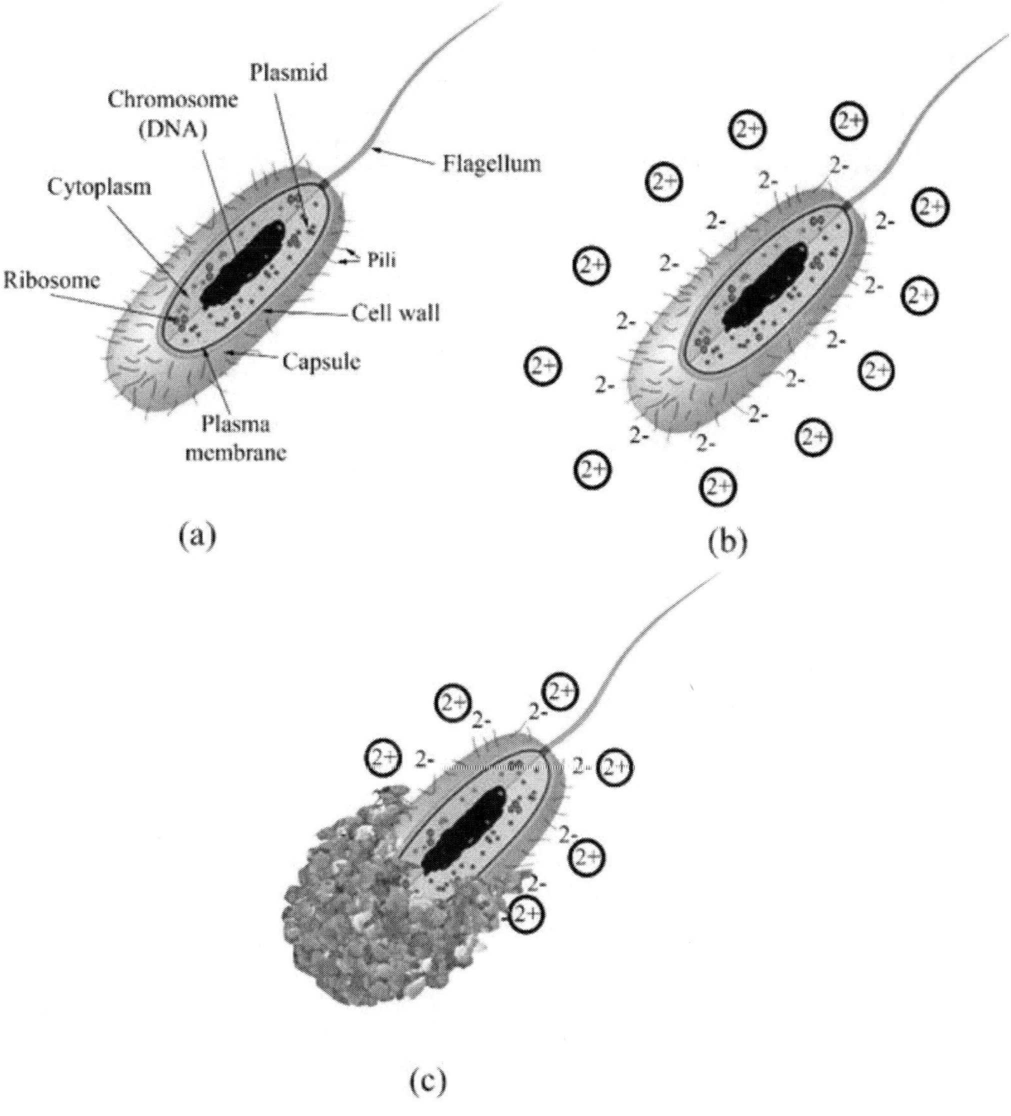

Figure 2. (a) Bacteria structure; (b) Negative charged cell wall and presence of positive charged ions; (c) Biomineral production by means of binding ions to cell wall.

Calcium carbonate deposition is basically considered a common phenomenon among bacterial activity if the calcium source as precursor exists in media [29]. Presence of carbonate and calcium ion in high alkaline conditions yields to calcium carbonate precipitation [30]. Consolidation of stone, removal of metals from industrial waste water streams and disposal, trace pollutants in ground water supply, soil strength improvement, soil erosion control, and production of minerals plugging in oil industry are the main implementation of calcium carbonate biomineralisation [31-35].

As shown in Figure 2, calcium carbonate particles are formed as positively charged calcium ions attached to the negatively a charged bacterial cell wall. The rate of calcium carbonate precipitation depends on several factors, such as concentration of dissolved inorganic carbon, the pH of the local site, the concentration of calcium ion, Hartree energy (Eh), and the presence of the nucleation site where precipitation could take place [36, 37].

BIOMINERALISATION IN CEMENTITIOUS MATERIALS

Recently, the biomineralisation process has been successfully implemented in construction materials, such as cementitious materials and limestone, to fill the cracks (Figure 3).

High pH (i.e., 11–13) and environmental conditions make the bacteria vulnerable to live in the concrete. They are therefore usually placed in concrete matrix through various ways, such as encapsulation, tubular network and mixing with other composition. As it can be seen from Figure 4a, bacteria and nutrients are mixed with concrete ingredients. However, to ensure variability of bacteria in the long term, they have been immobilised in protective particles, such as clay particles, diatomaceous earth and granular activated carbon [39, 40]. As the crack appears, bacteria germinate upon exposure to water and nutrients available in the concrete. Then precipitation of calcium carbonate prevents movement of aggressive materials and it also restricts crack developments.

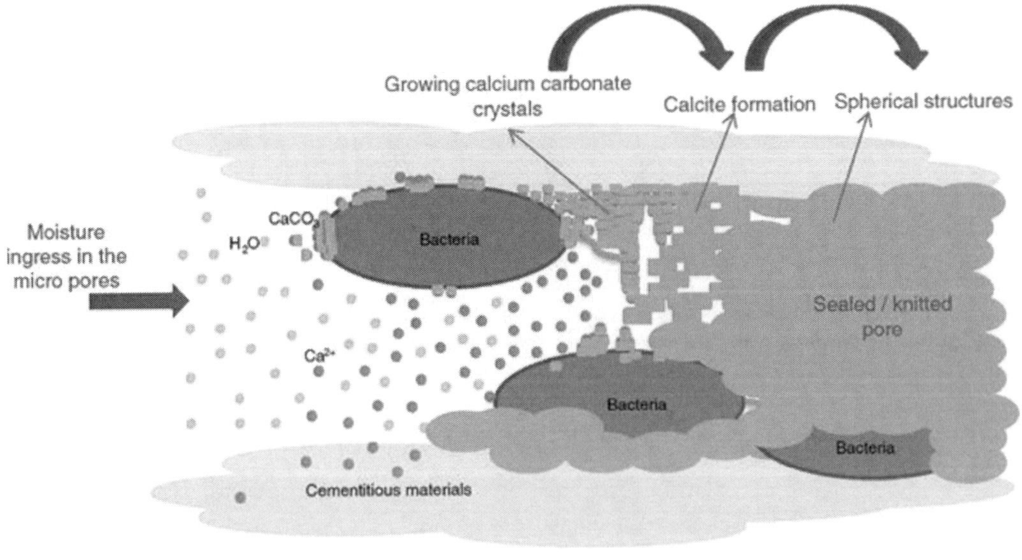

Figure 3. Activation of bacterial strain upon water movement through cracks [38].

On the other hand, capsule based healing relies on cylindrical or ball shaped capsules that are broken upon the generation of the crack. As seen in Figure 4b, bacteria and nutrients are encapsulated in microcapsules. Capsules are mixed with other ingredients during concrete preparation. Incorporation of the healing agent into microcapsules and also vulnerability of capsules during mixing are the main causes to restrict this pathway.

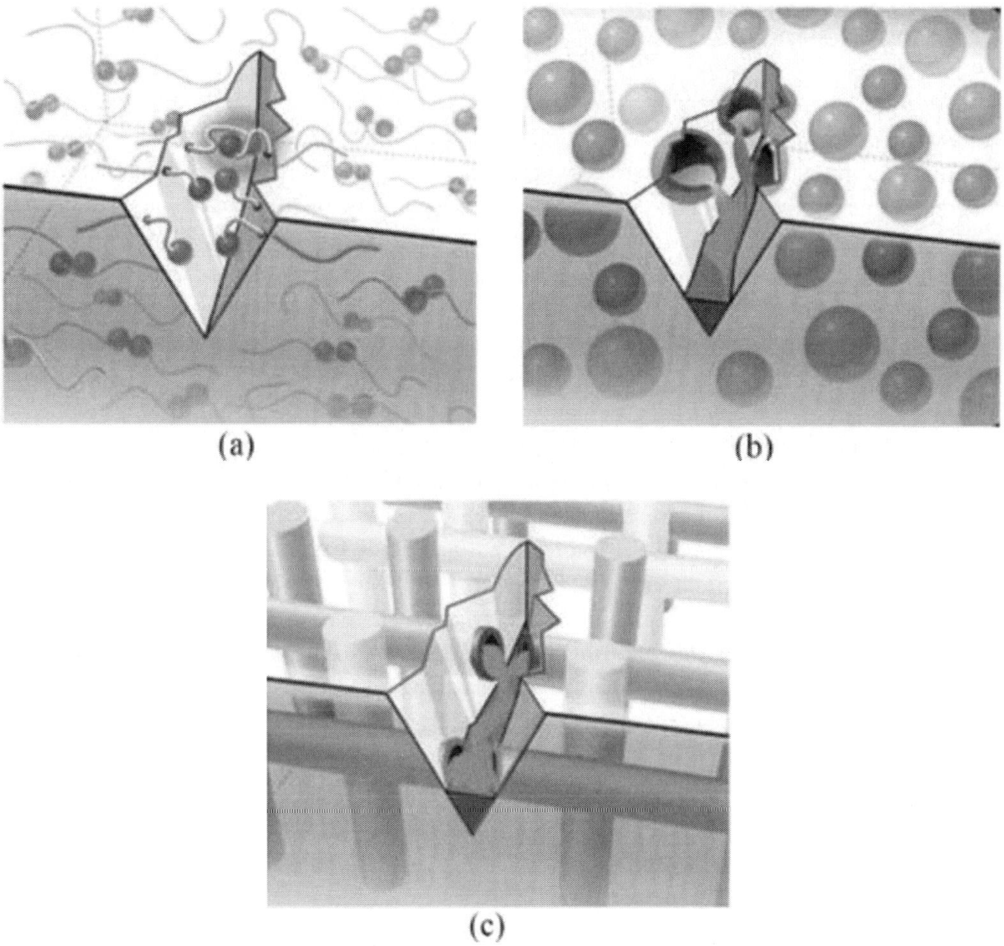

Figure 4. Three main self-healing types: (a) mixing with other ingredients (b) encapsulation (c) vascular [41].

The mechanism of activation in the vascular healing technique is the same as capsule based except the healing agent is kept in the vein that runs through the concrete instead of the capsule that sits in a discrete location. As can be depicted in Figure 4c, the healing agent is provided by an external reservoir which is placed outside of the matrix structure. Since the crack occurs, the healing agent runs through the vascular network due to pressure gradient between the reservoir which is located higher than the structure and crack vicinity. However, if the healing agent released more than crack capacity, it causes aesthetic issues and also increases the maintenance cost to compensate the healing agent lost. In addition, fracture of those tubes which are located closer to the surface may cause dramatic reduction in healing capacity. Since the healing agent released upon tube rupture, it may prevent the transfer of the healing agent to other parts.

Microbial Concrete Efficiency

Strength Assessment

Concrete strength capacity mostly refers to compressive and tensile strength. Compressive strength is one of the most significant attributes of concrete, which is obtained from maximum resistance of concrete to applied pressure. Compressive strength capacity of concrete is generally enhanced over time [42]. Depending on the applied load, different stresses, such as compression, tension, bending or flexural, shear and torsion, will be produced in concrete structures. Sample size and shape, aggregate size and type, curing condition, temperature, humidity, demoulding time, water to cement (w/c) ratio, utilisation of admixture and curing age have a profound impact on the compressive strength result [42-44]. According to the American Society for Testing and Materials (ASTM), C 109/C 109M, compressive strength of hydraulic cement mortar is determined by using 150-mm cubes at age of 3, 7 and 28 days [45]. As a result of shape effect on compressive strength, the cubic sample shows higher compressive strength than the cylindrical sample. To overcome this issue, correction factor has been suggested by ASTM provision when the height to diameter ratio is less than 2 [44]. Whereas the International Standards Organisation (ISO) 679 specifies the application of 40×40×160 mm prismatic sample with water to cement ratio of 0.5 [46].

Some microbial strains, such as *Bacillus sphaericus, Sporoscarcina pasteurii, Bacillus subtilis, Bacillus cohnii and Diaphorobacter nitroreducens,* have been scientifically proven to be able to produce calcium carbonate to fill the cracks [47-52]. It has been reported that incorporation of *Bacillus pseudofirmus* in cement mortar reduces compressive strength compared to reference sample [53]. Similarly, compressive test at the age of 90 days revealed that increasing the number of microcapsules (concentration of *Bacillus sphaericus*) incorporated into the microcapsules results to decrease compression capacity [8]. Moreover, effect of water to cement ratio and also curing time disclosed that the latter parameter had less influence on compressive strength [54].

On the contrary to compressive strength reduction, some investigations demonstrated improvement in compressive strength capacity. Different concentration of *Bacillus sp.* led to increased compressive strength up to 40% for aging from 7 to 28 days [47]. Likewise, 36% improvement in compressive strength was reported for 28 days incubated specimens which mixed with *Bacillus sp.* [55]. Bang et al. investigated the influence of immobilisation of bacterial strain on concrete properties [7, 56]. Although immobilisation of *S. pasteurii* in Siran^{TM} beads caused the increase in compressive strength for 7 and 28 days, immobilisation bacterial strain in prepolymer of hydrophilic polyurethane led to decrease the compressive strength for the same period of time. In order to measure concrete tensile strength, ASTM has provided two types of flexural test: ASTM C 78 – 94 for using simple beam with third-point loading and ASTM C 293 – 94 for utilisation of simple beam with center-point loading [57, 58]. Although it is reported that immobilised bacteria in microcapsules did not have the significant impact on tensile strength, the increase in the number of microcapsules led to decrease tensile strength [8].

Durability Assessment

Apart from concrete mechanical properties, durability has profound influence on service life of structures. The outer part of concrete (virtually 30-50 mm) plays the more crucial role in reduction of durability than the inner part [54]. For instance, pavement concrete lifespan can be substantially decreased upon exposure of aggressive materials and dynamics load which is largely produced by vehicles [59]. Aggressive movement from the environment into concrete yields to interaction with concrete constituents, corrosion of reinforcement bar and consequently it effects the integrity of the concrete structure [60].

Durability of concrete is evaluated through various transportation mechanisms in respect to the driving force: permeability, water absorption and diffusion [61]. Structural attack by means of sulphate, sea water and acid are the major reasons of structural degradation. Deterioration level relies on aggressive concentrations and the permeability of the concrete. It has been noted that durability of concrete can be enhanced by adding chemicals and mineral admixtures, application of special reinforcement bars, installation of cathodic and external coating [62]. Despite contradictory impact of bacteria on compressive strength, results indicate that incorporation of bacterial strain has a positive effect on the durability of the concrete structure. The following sections describe the different transportation mechanisms of liquid or gas into concrete structures.

Permeability

Egress resistance of liquids or gases to the concrete is defined as permeability. Theoretically permeability refers to fluid flow rate egress under pressure in saturated medium, whereas water absorption indicates the water movement into concrete through capillary suction [61, 63]. Permeation of concrete mostly depends on pore size, its distribution, water/cement ratio and concrete age. Although more capillary pores in the structure results in an increase permeability coefficient, and reduction in water to cement ratio leads to reduce permeation. Water, gas and rapid chloride permeability are main tests to evaluate concrete permeation [64].

As seen from Eq. 6, permeability of concrete is calculated by Darcy's expression. Where Q is flow rate, A is cross sectional area, K is permeability coefficient, dh is head loss, and dL is the sample flow path thickness. It shows fluid rate penetration in steady-state condition is proportional to hydraulic gradient. Hydraulic conductivity (constant coefficient) depends on porosity, the fluid properties moving through concrete, and the degree of concrete homogeneity [60, 65]. The higher permeability paves the path for the faster transportation of aggressive fluids or gasses and therefore it results to accelerate concrete degradation [59].

$$v = \frac{Q}{A} = -k(\frac{dh}{dL}) \tag{6}$$

A practical laboratory permeation test can be conducted as all specimen's faces are sealed except the top face which is subjected to pressured water. Since water penetration reaches to steady state, amount of loss water in head tank in the given time results to calculate permeability coefficient. In the case of using viscous fluids, generated form of Darcy's expression can be applied where μ is fluid viscosity (Eq. 7).

$$v = \frac{Q}{A} = -(\frac{k}{\mu})(\frac{dh}{dL}) \tag{7}$$

Due to the importance of permeability, a permeation test has been carried out before and after crack treatment. As it has been claimed, application of *Bacillus sphaericus* immobilised in hydrogel decreased water permeability by 68% in average, while permeability in non-bacterial specimens was between 15-55% [66]. In another investigation, the same author found that melamine microcapsules containing *Bacillus sphaericus* can reduce permeability up to 10 times compared to free cell samples [8]. Van Tittelboom et al. [21] observed that live and dead *Bacillus sphaericus* immobilised in silica sol gel were able to decrease water permeability. It was concluded that further permeability reduction in live series attributes to precipitation of calcium carbonate crystals. Comparison between bacteria immobilised in silica gel and polyurethane revealed that the more decrease in permeation occurred in the latter condition. However, in both cases permeability coefficient decreased compared to reference samples.

Water Absorption

Capillary water absorption is a phenomenon where fluids move in porous solids and it is determined by two methods, namely surface and drilled hole absorptivity test. In the first case, penetration of water through surface is considered; however, the latter one, water absorption, is determined through the drilled hole [60]. The standard test for measurement of water absorption has been prescribed by ASTM. The test is based on measurement of increase mass in concrete disc at regular time intervals [67]. To provide flow movement in one direction, the disc circumference should be coated with waterproof materials.

The modified capillary water absorption test, according to RILEM 25 PEM II-6, has been performed, where the water absorption coefficient is determined at regular time intervals by Eq. 8; where Q is absorbed water weight, S is the uncoated area, and t is the time. Wang et al. [40] demonstrated the application of immobilised bacteria into diatomaceous earth can reduce water uptake; however, no considerable difference was observed between the samples with no bacteria which were immersed in water or deposition medium [40].

$$\frac{Q}{S} = k\sqrt{t} \tag{8}$$

The deposition of calcium carbonate layer as a consequence of bacterial activity results in decreased sorptivity. It is found that that *Bacillus sp.* can substantially decrease water absorption in concrete structures [55]. It was reported that the reduction in water to cement ratio in the same curing durations led to a decrease in water absorption value. Reduction of water to cement ratio implies less porosity in concrete and consequently denser material [54]. Furthermore, the effect of *Sporoscarcina pasteurii* with fly ash in concrete was investigated. It is noted that presence of bacteria can decrease water absorption in specimens with or without fly ash [48]. In comparison with bacteria incorporated in ordinary concrete samples, existence of bacteria with fly ash substantially reduced water absorption.

Gas Permeability

The gas permeability generally depends on pore size, curing duration, saturation degree, testing age, distribution of moisture, applied pressure and water to cement ratio [54, 61]. Although gas permeation is a relatively simple test, it takes a long time due to measurement in different ages. It has been observed that the increase in curing time and also reduction of water to cement ratio lead to a decrease in gas permeation values. It has been concluded that gas permeability and water absorption values are linearly proportional to one another [54]. While sensitivity of gas permeation to duration of curing sample and w/c ratio is more than water absorption.

Chloride Ion Penetration

Concentration gradient leads to movement of matter throughout the specimen which is called diffusion. Diffusion in gasses is faster than liquids and solids. The concrete diffusion test is broadly divided to three main groups: ionic diffusion, gas diffusion and vapour diffusion [60]. Steady state, non-steady state and electric field migration tests are the common ionic diffusion tests [60]. The use of electrical field in an ionic migration test was suggested by Whiting [68]. It caused a significant reduction in experiment duration, while steady state and non-steady state tests required considerable involvement of time.

Several investigations have been conducted to evaluate chloride ion penetration in bacterial self-healing concrete. Although chloride ion attack has no direct influence on concrete degradation, its transportation has a destructive impact on acceleration of reinforcement bar corrosion. In this case the formation of rust as a result of chloride ionic attack paves the path for spalling, crack formation and concrete delamination [11]. Rapid chloride permeation is carried out by prescribed procedure in ASTM in specific interval times [69]. It was reported that utilisation of bacteria up to 10^5 cells/ml decreased rapid chloride permeability in presence of fly ash, whereas concentration of bacteria at 10^7 cells/ml resulted in increased chloride permeability [48]. The improvement of durability verifies that a biological approach can fulfil the requirements for being a self-healing mechanism.

Pore Size

Pore size and its distribution have considerable effect on durability. It was concluded that the porosity, pore size and its distribution are influenced by water to cement ratio and cement particle hydration [70]. The results indicated that the application of lower w/c ratio, the increase of curing duration led to decreased pore size and consequently porosity [71, 72]. However, water to cement ratio and cement fineness have a greater impact on porosity than curing duration [73]. Furthermore, existence of more porosity in Portland cement than blended cement in the same curing condition and w/c ratio confirms influence of concrete composition on porosity [73].

Mercury Intrusion Porosimeter (MIP) has been widely used to determine pore size and its distribution across concrete specimens. The specimen is placed in a MIP chamber and pressure is increased after evacuation of the chamber. Tracking pressures and intrusion volumes result in measure connecting pore necks and breakthrough pressure in continuous and discontinuous systems respectively [74]. Real porosity values would be greater than MIP results if pores were too isolated or small. As it has been reported, concrete age has an effect on the pore size. MIP analysis demonstrated bigger pore size in young specimens with the age of 3 to 7 days than older ones (28 days cured) [53].

Wang et al. [8] studied the effect of incorporated bacteria on porosity of concrete. It was observed that pore size in the samples accompanied by nutrients, microcapsules and bacteria have lower pore size (mostly around 0.04-0.05 μm) than those without bacteria and microcapsules (mostly around 0.1 μm).

Calcium Carbonate Measurement and Morphology

Apart from calcium carbonate production through biological activity, quantitative determination as well as morphological analysis of product are important in both natural and synthetic conditions. Due to the existence of different polymorphic in precipitated calcium carbonate and their significant role in biomineralisation, discrimination of them has become the aim of many studies.

As depicted in Figure 5, calcium carbonate has three anhydrous crystalline forms, namely calcite, aragonite and vaterite. Calcite (rhombohedra crystal) is known as the most thermodynamically stable at an ambient temperature and best characterised polymorph of calcium carbonate. It has been noted that low carbon dioxide pressure results in calcite precipitation. Aragonite (the needle-like crystal) is the least abundant calcium carbonate crystalline and it mostly produces through biomineralisation [75]. It is denser than calcite, and therefore accumulation of aragonite particles in cracks may cause to enhance concrete integrity by increasing the bonding capacity. Vaterite, (hexagonal crystal) as a meta-stable polymorph of calcium carbonate, is the most hydrophilic crystal in comparison to the other two crystalline polymorph [76]. Precipitated vaterite particles are the least stable form of calcium carbonate and mostly is produced at a high level of supersaturation. Therefore, based on the temperature and pH, vaterite serves as a precursor phase of other polymorphs.

Production of calcium carbonate has been quantified through various techniques, such as Thermogravimetric Analysis (TGA), Fourier Transform Infrared Spectroscopy (FT-IR), Raman Spectroscopy and X-ray diffraction. Although vibrational methodologies are able to discriminate different crystal polymorphs in ternary mixtures, they are unable to quantify these crystals due to overlapping peak bands [77]. The following sections will describe the relevance of each technique to determination of calcium carbonate.

Characterisation of Calcium Carbonate

Thermogravimetric Analysis

There are different approaches for thermal analysis: thermogravimetric analysis (TGA), differential scanning calorimetry, differential thermal analysis, pressurised thermogravimetric analysis, thermo mechanical analysis, dilatometry and evolved gas analysis. They vary from each other in respect to the measured parameters. A common method to determine presence of compound in the sample is thermogravimetric analysis. TGA can be used for determination of reaction kinetics, material characterisation, and existence of organic and inorganic substances in unknown mixture. It is based on mass changes of composition upon gas adsorption, gas reaction, phase transitions, gas desorption, decomposition, break down reaction and chemisorption as temperature increases [78]. The unknown composition is heated continuously through the following ways: constant rate, isothermic and gradually isothermic. Mass changes are regularly measured as a function of temperature gradient. Therefore temperature-mass loss or gain history can determine the presence of organic contents in the sample as well as thermal stability and kinetic parameters of reactions. A number of studies have reported the presence of calcium carbonate by TGA [21, 79-81].

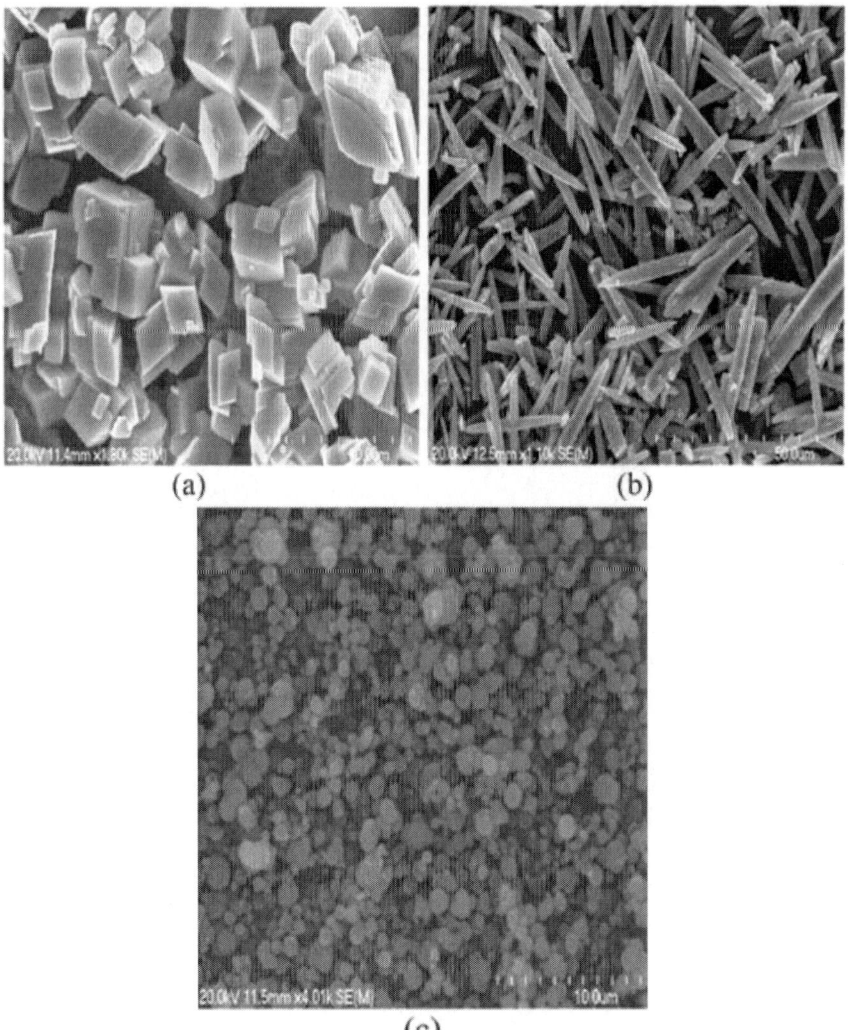

Figure 5. Calcium carbonate polymorphs: (a) Calcite; (b) Aragonite; (c) Vaterite.

As seen in Eq. 9, calcium carbonate decomposes to calcium oxide and carbon dioxide between 530-950°C. The composition weight loss due to dissipation of carbon dioxide during decomposition leads to quantify calcium carbonate [82].

$$CaCO_3 \rightarrow CaO + CO_2 \quad (9)$$

Reaction of Calcium Carbonate with Hydrochloride Acid

Although calcium carbonate is insoluble in water, it dissolves in hydrochloric acid (HCl). Therefore a common practice to quantify calcium carbonate is reaction with hydrochloric acid according to Eq. 10.

$$2HCl(aq.) + CaCO_3(s) \rightarrow Ca^{2+}(aq.) + CO_2(g.) + H_2O + 2Cl^-(aq.) \quad (10)$$

To ensure a complete reaction and also acceleration of reaction close to the endpoint, additional HCl is mostly added. Calcium content in the composition is estimated with the aim of loss in mass due to presence of carbon dioxide in Eq. 10.

Another approach is determination of the reacted HCl moles to quantify calcium carbonate. To calculate amount of excessive HCl, back titration with sodium hydroxide according to the Eq. 11 can be applied and consequently calcium content is measured from the mole of reacted HCl.

$$HCl(aq.) + NaOH(aq.) \rightarrow H_2O + Na^+(aq.) + Cl^-(aq.) \quad (11)$$

Self-Healing Efficiency through Optical Microscope

Bio self-healing efficiency can also be obtained according to reduction of crack width via light microscope during concrete remediation. Crack width is regularly monitored to determine efficiency of self-healing as well as bacterial activity. For this purpose some crack areas are marked, and therefore the crack width decreases under different intervals are recorded. The healing ratio is defined by Eq. 12 where C_{wi} is initial crack width and C_{wf} is final crack width [83].

$$r = [\frac{(C_{wi} - C_{wf})}{C_{wi}}] \times 100 \quad (12)$$

It is reported that the incorporation of the healing agent (*Bacillus alkalinitrilicus* and nutrients) into concrete matrix can heal 100% of the crack (for crack width less than 0.64 mm) during 100 days [84]. Other research demonstrated application of *Bacillus sphaericus* subjected to wet-dry cycles in water incubation led to fill crack width 970 μm which was four-fold greater than those samples of free bacteria [8]. However, healing efficiency in bio hydrogels incorporated specimens with a maximum crack width of 300 μm (*Bacillus sphaericus*) were in the range of 70-100%. Luo et al. [85] explored the efficiency of bio self-healing with the aid of an image processing technique. In the first step, crack intensity was thresholded by 115. Then the number of pixels on the crack were accumulated before and after healing, and therefore the repair rate of the crack area was evaluated. Figure 6 shows the crack filling capacity from initial measurement to 20 days healing.

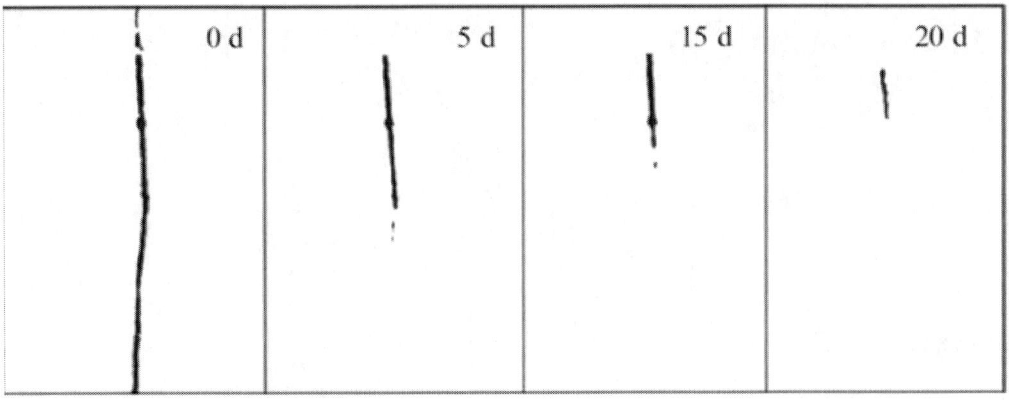

Figure 6. Crack filling capacity up to 20 days [85].

It should be noted that, in spite of calcium carbonate quantification, it is not easy to distinguish the portion of calcium carbonate which is produced through biological self-healing approach, so that, additional calcium carbonate is simultaneously produced from hydration of un-hydrated cement particles. However, precipitation of calcium carbonate through biological activity can be confirmed through different techniques.

Morphological Observation

Scanning Electron Microscope

Despite the fact that relatively new technologies, such as Scanning Electron Microscope (SEM), Energy Dispersive Spectrum (EDS), X-ray diffraction (XRD), Fourier Transform Infrared Spectroscopy (FT-IR) and similar instruments, have not been standardised in the field of civil engineering, their application in concrete technology are quickly developing [86].

SEM, as the most widely employed technique, would appear to be an ideal instrument to identify morphology of materials. To ensure that precipitated product in biological self-healing concrete is calcium carbonate, SEM has been employed by many scholars [7, 8, 29, 40, 53, 66, 87, 88]. Calcium carbonate particles produced by *Bacillus sphaericus and Bacillus subtilis* through BIM is shown in Figure 7. Despite obtaining comprehensive information from SEM images, estimation of precipitated calcium carbonate in order to calculate productivity of biological self-healing is impossible.

Energy Dispersive Spectrum

In most cases SEM is equipped with Energy Dispersive Spectrum (EDS) for chemical analysis. Determination of existence elements in composition are obtained with the aim of EDS analysis. A range of different results have been reported in respect to utilisation of bacteria, nutrients and curing age. Considering the element weigh ratio and the comparison with calcium carbonate ions (ca: c: o = 1:1:3), it can be concluded from Figure 8 that most of the present particles are calcium carbonate.

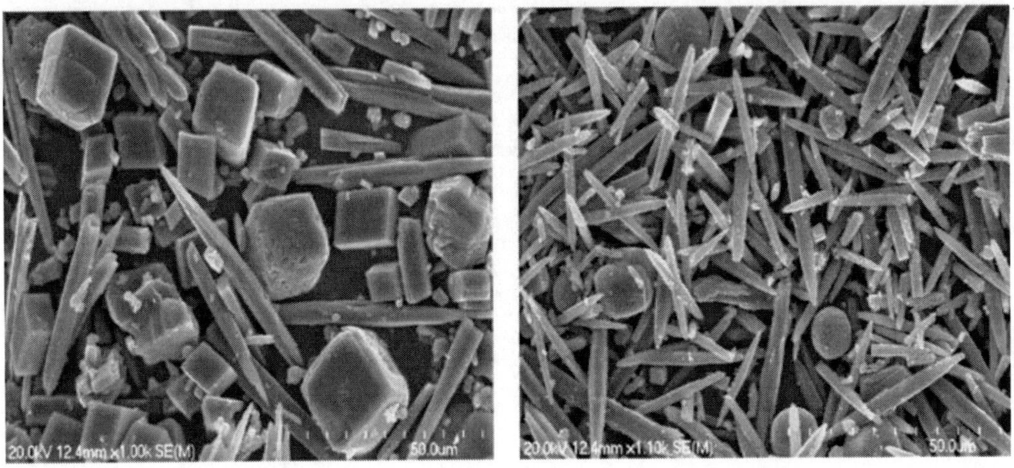

(a) (b)

Figure 7. (a) SEM micrographs of calcite precipitation by *B. sphaericus*; (b) SEM micrographs of calcite precipitation by *B. subtilis*.

Although presence of calcium, carbon and oxygen can be proven from EDS analysis, existence of calcium does not exactly imply the formed calcium carbonate. Homogenously distribution of calcium, carbon and oxygen in elemental mapping makes it impossible to distinguish combination of these elements (Figure 9). Furthermore, due to presence of unreacted calcium ions, it cannot be concluded that the whole of calcium ions existed in the EDS result represented by precipitated calcium carbonate.

Fourier Transform Infrared Spectroscopy and Raman Spectroscopy

Fourier Transform Infrared Spectroscopy (FT-IR) is used for identification of unknown materials, molecular structures and quantitative determination of components in a mixture through absorption and transmission of infrared radiation. It is able to analyse a small amount of a wide range of samples, such as powder, solid, liquid and thin film. The peak size indicates the amount of existing elements in the compound. Despite its advantages, this technique is unable to determine polymorph contents. Moreover, vaterite and aragonite may convert to calcite during pelling and grinding; the sample also needs to be mixed with KBr [77, 89, 90].

Due to overlapping band over three polymorphs, application of infrared spectrometry has become restricted in ternary phases [89, 90]. To overcome absorption overlapping band in ternary mixture, Vagenas et al. [89] proposed a model to identify calcium carbonate polymorphs (calcite, vaterite, arogonite) with the aim of Beer's law.

Raman spectroscopy is another technique which has been widely used to quantitate the amount of polymorphic crystals in binary compositions. Raman spectroscopy is a non-absorption technique which is based on light scattering. Similar to FT-IR, it is unable to discriminate ternary mixtures. However, the application of a new technique has been suggested to identify polymorphs of calcium carbonate in ternary mixtures [90, 91]. Observation of calcite and aragonite crystals in the crack zone, which are akin to the standard calcium carbonate crystals, has confirmed the possibility of calcium carbonate precipitation in concrete through biological approach [84].

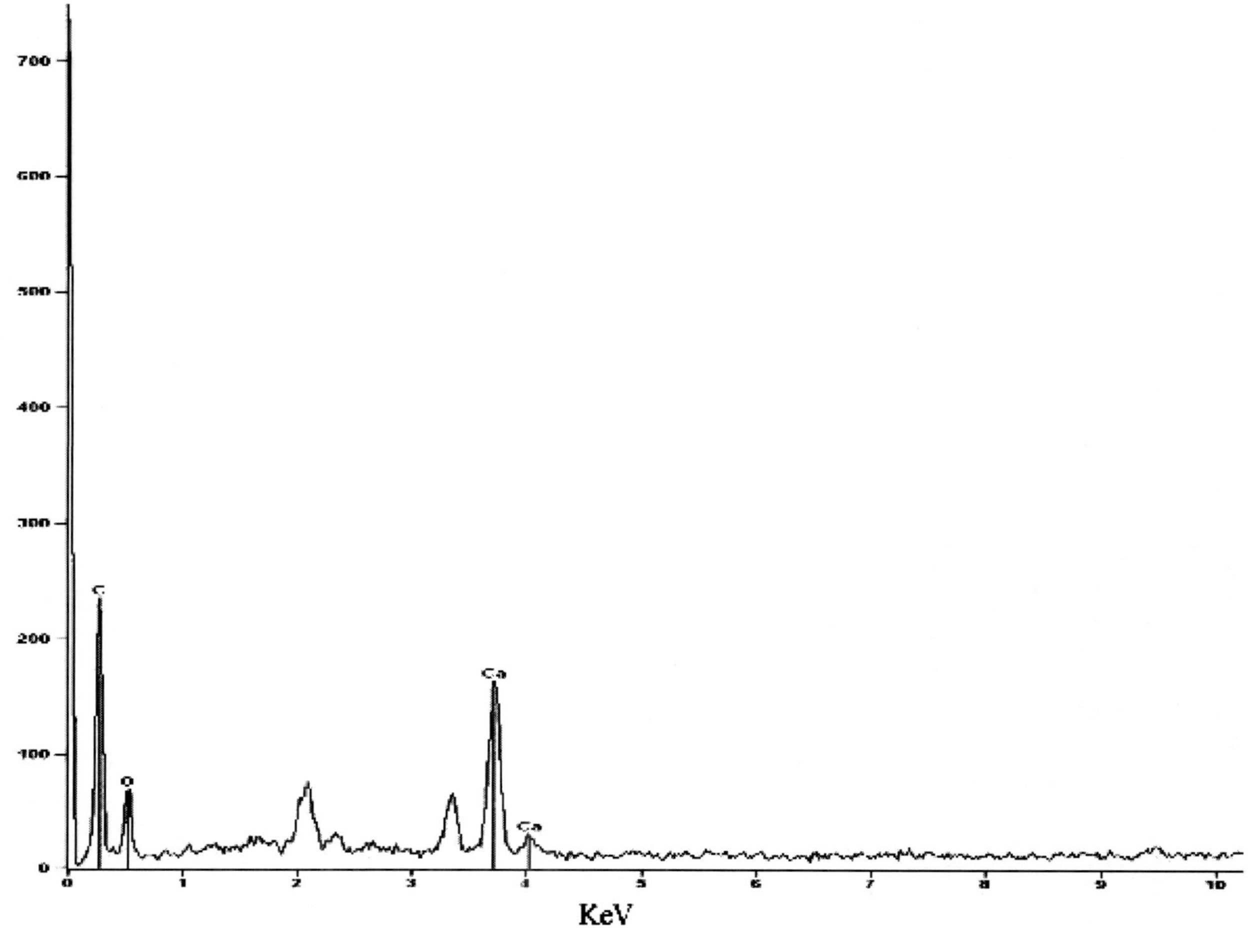

Figure 8. EDS analysis of produced calcium carbonate.

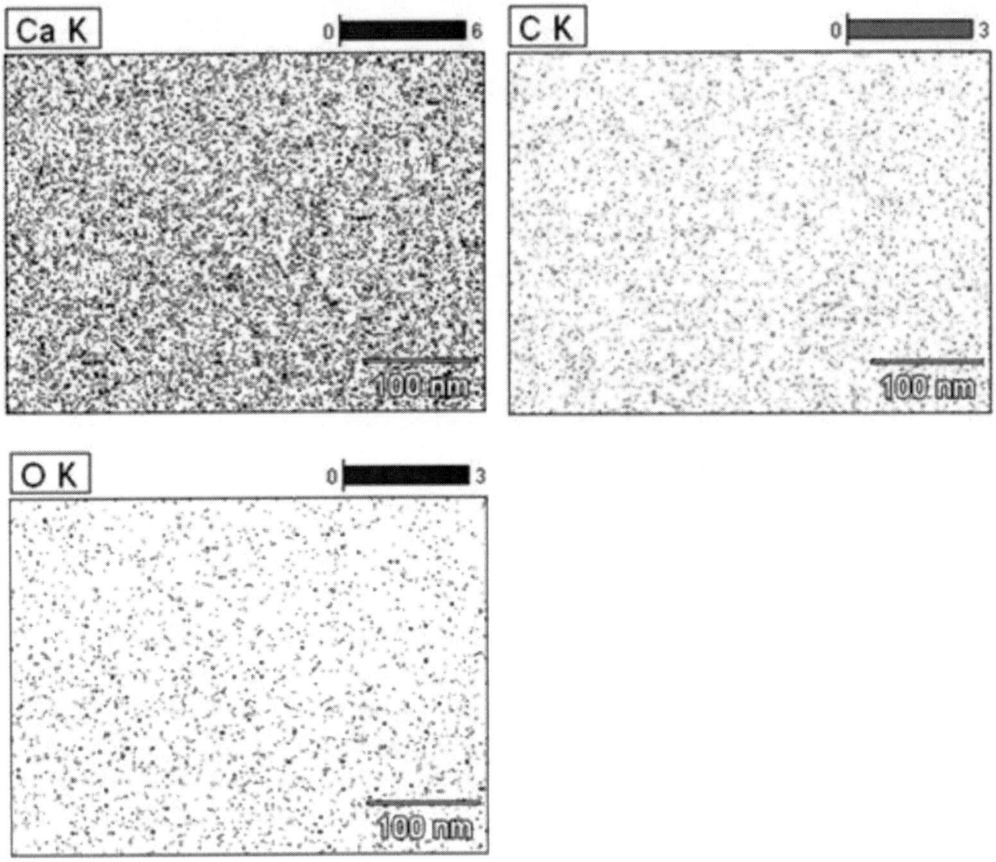

Figure 9. Corresponded EDS elemental mapping images of produced calcium carbonate.

X-Ray Diffraction and X-Ray CT

X-ray diffraction as a non-destructive analytical characterisation technique are employed to identify crystalline phases that may present in materials [92]. Data is collected from a different range of degrees and is then compared to databases to phase identification and quantitative analysis. Each pattern is representative of a specific element in composition due to a unique pattern for each substance. It provides compositional and structural information about crystal structure which is distinguishable from diffraction patterns, crystal size and crystalline orientation.

XRD have been used to determine crystallographic calcium carbonate structure as well as crystal orientation in the field of self-healing concrete. Comparison of bio-based concrete samples which contained *S. pasteurii* and *B. sphaericus* demonstrates more of a tendency towards precipitation of vaterite and calcite than aragonite [88]. While Kim et al. [85] reported the presence of calcite in the healed crack through biological activity.

The quality and quantity of bio base healing approach must be considered for the cracks and pores located inside and on the surface of concrete structures. Although the majority of researchers have been focused on the possibility of calcium carbonate precipitation on the surface of cementitious materials, X-ray tomography has been used to observe the possibility of calcium carbonate production inside the samples [83]. In spite of the fact that X-ray CT can detect calcium carbonate precipitation inside the matrix as well as the surface, dependency on sample size and the lack of elemental composition detection are the main restrictions of using this method.

Results indicated that bacteria incorporated in polymeric hydrogels are able to produce calcium carbonate inside the concrete matrix. Presence of more oxygen and carbon dioxide on the surface of the concrete structures leads to an increase in the efficiency of bio self-healing on the surface compared to the interior part of the concrete matrix. Overall, due to the shortcomings associated with Infrared spectroscopy and Raman spectroscopy, quantitative detection of polymorphs via X-ray diffraction is superior to Infrared spectroscopy and Raman spectroscopy.

CONCLUSION AND FUTURE TRENDS

Apart from concrete robustness via bio based healing approach, the cost of producing bio based healing concrete has remained a matter of debate. Bacteria, nutrients and labour are the major costs to provide bio self-healing agent. As it was reported by Muynck et al. [93], the cost for the preparation of lyophilized bacteria and nutrients are about $1240 and $203 per kilogram, respectively. It was found that rehabilitation of typical concrete surface (1×1m) is $26–32 considering the need of 2-3 gr/m^2 bacteria and 0.04-0.08 kg/m^2 nutrients. However, this amount will be increased by $40–45 for serious damaged surfaces. It is suggested that utilisation of alternative nutrients from industrial waste may reduce costs. Studies demonstrated that utilisation of alternative nutrient sources, such as lactose mother liquor and corn steep liquor (CSL), could increase concrete mechanical properties [94, 95]. Based on the price of nutrients, using CSL instead of standard laboratory nutrients may decrease the cost of biological self-healing by three times. Moreover, results indicated the improvement in permeability as well as calcium carbonate precipitation efficiency in the presence of CSL. Moreover, microorganism effectiveness highly relies on environmental condition. Hence, proper immobilisations, which are compatible with matrix and are also tolerant in harsh environments, require more investigation.

The great success of bio self-healing concrete depends on the improvement in both mechanical properties and durability. The decrease in permeability is mainly related to the existence of newly formed calcium carbonate as a result of biomineralisation. In contrast, the contradictory results for compressive strength may attribute to low binding capacity and brittleness of the bio product. Therefore to encourage contractors to use bio self-healing concrete, a great amount of emphasis should be placed on addressing the negative effects of bio based agents on compressive strength and also reducing the cost as reasonably achievable.

REFERENCES

[1] Neville, A. M; Brooks, J. J. (2010). Concrete Technology. 2nd Edition.
[2] Worrell, E; Price, L; Martin, N; Hendriks, C; Meida, L. O. (2001). Carbon dioxide emissions from the global cement industry. *Annu. Rev. Energy Environ.*

[3] Toma, I. O; Taranu, G; Toma, A. M; Budescu, M. (2011). Influence of cement and sand type on the strength characteristics of mortars with various contents of green binder.

[4] Statista. (2015). United States and world cement production in 2010 and 2014.

[5] The Washington Post. (2014). How China used more cement in 3 years than the U. S. did in the entire 20th Century.

[6] Mohiuddin, A. K. (2010). Concrete Repair Methods, in Bridge and Highway Structure Rehabilitation and Repair. McGraw Hill Professional. Access Engineering.

[7] Bang, S. S; Lippert, J. J; Yerra, U; Mulukutla, S; Ramakrishnan, V. (2010). Microbial calcite, a bio-based smart nanomaterial in concrete remediation. *International Journal of Smart and Nano Materials*. 1, 28-39.

[8] Wang, J. Y; Soens, H; Verstraete, W; De Belie, N. (2014). Self-healing concrete by use of microencapsulated bacterial spores. *Cement and Concrete Research*. 56,139-152.

[9] Collins, F; Sanjayan, J. G. (2000). Effect of pore size distribution on drying shrinkage of alkali-activated slag concrete. *Cement and Concrete Research*. 30, 1401-1406.

[10] Claisse, P. A; Elsayad, H. I; Shaaban, I. G. (1997). Absorption and sorptivity of cover concrete. *Journal of Materials in Civil Engineering*. 9, 105-110.

[11] Ababneh, A; Benboudjema, F; Xi, Y. (2003). Chloride penetration in nonsaturated concrete. *Journal of Materials in Civil Engineering*. 15, 183-191.

[12] Edvardsen, C. (1999). Water permeability and autogenous healing of cracks in concrete. *ACI Materials* Journal. 96, 448-454.

[13] Hearn, N. (1998). Self-sealing, autogenous healing and continued hydration: What is the difference? *Materials and Structures/Materiaux et Constructions*. 31, 563-567.

[14] Şahmaran, M; Keskin, S. B; Ozerkan, G; Yaman, I. O. (2008). Self-healing of mechanically-loaded self consolidating concretes with high volumes of fly ash. *Cement and Concrete Composites*. 30, 872-879.

[15] Reinhardt, H. W; Jooss, M. (2003). Permeability and self-healing of cracked concrete as a function of temperature and crack width. *Cement and Concrete Research*. 33, 981-985.

[16] Clear, C. A. (1985). Effects of autogenou healing upon the leakage of water through cracks in concrete in Technical Report - *Cement and Concrete Association*.

[17] Van der Zwaag, S. (2007). Self Healing Materials: An Alternative Approach to 20 Centuries of Materials Science. Springer Netherlands.

[18] Jonkers, H. M; Schlangen, E. (2009). A two component bacteria-based self-healing concrete. Proceedings of the 2nd International Conference on Concrete Repair, Rehabilitation and Retrofitting, ICCRRR.

[19] De Muynck, W; Debrouwer, D; De Belie, N; Verstraete, W. (2008). Bacterial carbonate precipitation improves the durability of cementitious materials. *Cement and Concrete Research*. 38, 1005-1014.

[20] Dhami, N.; Mukherjee, A; Reddy, M. S. (2012). Biofilm and microbial applications in biomineralized concrete.

[21] Van Tittelboom, K; De Belie, N; De Muynck, W; Verstraete, W. (2010). Use of bacteria to repair cracks in concrete. *Cement and Concrete Research*. 40, 157-166.

[22] Dry, C. (1994). Matrix cracking repair and filling using active and passive modes for smart timed release of chemicals from fibers into cement matrices. *Smart Materials and Structures*. 3, 118-123.

[23] Boyjoo, Y; Pareek, V. K; Liu, J. (2014). Synthesis of micro and nano-sized calcium carbonate particles and their applications. *Journal of Materials Chemistry A*. 2, 14270-14288.

[24] Santos, R. M; Ceulemans, P; Van Gerven, T. (2012). Synthesis of pure aragonite by sonochemical mineral carbonation. *Chemical Engineering Research and Design*. 90, 715-725.

[25] Hu, Z; Deng, Y. (2004). Synthesis of needle-like aragonite from calcium chloride and sparingly soluble magnesium carbonate. *Powder Technology*. 140, 10-16.

[26] Tebo, B. M; Johnson, H. A; McCarthy, J. K; Templeton, A. S. (2005). Geomicrobiology of manganese(II) oxidation. *Trends in Microbiology*. 13, 421-428.

[27] Mann, S. (2001). Biomineralization: Principles and concepts in bioinorganic materials chemistry. 6th ed. New York, Oxford University Press.

[28] Rivadeneyra, M. A; Delgado, R; del Moral, A; Ferrer, M. R; Ramos-Cormenzana, A. (1994). Precipatation of calcium carbonate by Vibrio spp. from an inland saltern. *FEMS Microbiology Ecology*. 13, 197-204.

[29] Boquet, E; Boronate, A; Boronate, A. (1973). Production of calcite (calcium carbonate) crystals by soil bacteria is a general phenomenon. *Nature*. 246, 527 - 529.

[30] Rodriguez-Navarro, C; Rodriguez-Gallego, M; Chekroun, K. B; Gonzalez-Muñoz, M. T. (2003). Conservation of ornamental stone by Myxococcus xanthus-induced carbonate biomineralization. *Applied and Environmental Microbiology*. 69, 2182-2193.

[31] Dhami, N. K; Reddy, M. S; Mukherjee, A. (2013). Bacillus megaterium mediated mineralization of calcium carbonate as biogenic surface treatment of green building materials. *World Journal of Microbiology and Biotechnology*. 29, 2397-2406.

[32] De Muynck, W; Cox, K; Belie, N. D; Verstraete, W. (2008). Bacterial carbonate precipitation as an alternative surface treatment for concrete. *Construction and Building Materials*. 22, 875-885.

[33] Diels, L; Van Roy, S; Somers, K; Willems, I; Doyen, W; Mergeay, M; Springael, D; Leysen, R. (1995). The use of bacteria immobilized in tubular membrane reactors for heavy metal recovery and degradation of chlorinated aromatics. *Journal of Membrane Science*. 100, 249-258.

[34] Burbank, M. (2010). Precipitation of Calcite by Indigenous Microorganisms to Strengthen Soils. University of Idaho.

[35] Whiffin, V. S; van Paassen, L. A; Harkes, M. P. (2007). Microbial carbonate precipitation as a soil improvement technique. *Geomicrobiology Journal*. 24, 417-423.

[36] Barton, L. L; Northup, D. E. (2011). Microbes at Work in Nature: Biomineralization and Microbial Weathering. Hoboken, NJ, USA: in Microbial Ecology, John Wiley and Sons, Inc.

[37] Hammes, F; Verstraete, W. (2002). Key roles of pH and calcium metabolism in microbial carbonate precipitation.

[38] Sarayu, K; Iyer, N. R; Murthy, A. R. (2014). Exploration on the biotechnological aspect of the ureolytic bacteria for the production of the cementitious materials - A review. *Applied Biochemistry and Biotechnology*. 172, 2308-2323.

[39] Erşan, Y. Ç; Da Silva, F. B; Boon, N; Verstraete, W; De Belie, N. (2015). Screening of bacteria and concrete compatible protection materials. *Construction and Building Materials*. 88, 196-203.

[40] Wang, J. Y; De Belie, N; Verstraete, W. (2012). Diatomaceous earth as a protective vehicle for bacteria applied for self-healing concrete. *Journal of Industrial Microbiology and Biotechnology*. 39, 567-577.

[41] Blaiszik, B. J; Kramer, S. L. B; Olugebefola, S. C; Moore, J. S; Sottos, N. R; White, S. R. (2010). *Self-healing polymers and composites, in Annual Review of Materials Research*. 179-211.

[42] Silva, R. V; Brito, J; Dhir, R. K. (2015). The influence of the use of recycled aggregates on the compressive strength of concrete: a review. *European Journal of Environmental and Civil Engineering*. 19, 825-849.

[43] Sudin, M. A. S; Ramli, M. (2014). Effect of specimen shape and size on the compressive strength of foamed concrete. in MATEC Web of Conferences.

[44] Sim, J. I; Yang, K. H; Kim, H. Y; Choi, B. J. (2013). Size and shape effects on compressive strength of lightweight concrete. *Construction and Building Materials*. 38, 854-864.

[45] ASTM C109 / C109M-13. (2013). Standard Test Method for Compressive Strength of Hydraulic Cement Mortars (Using 2-in. or [50-mm] Cube Specimens). ASTM Internationall, West Conshohocken, PA.

[46] International Standards Organisation, ISO 679. (2009). Cement (Test methods) Determination of strength.

[47] Achal, V; Mukerjee, A; Sudhakara Reddy, M. (2013). Biogenic treatment improves the durability and remediates the cracks of concrete structures. *Construction and Building Materials*. 48, 1-5.

[48] Chahal, N; Siddique, R; Rajor, A. (2012). Influence of bacteria on the compressive strength, water absorption and rapid chloride permeability of fly ash concrete. *Construction and Building Materials*. 28, 351-356.

[49] Tiano, P; Biagiotti, L; Mastromei, G. (1999). Bacterial bio-mediated calcite precipitation for monumental stones conservation: Methods of evaluation. *Journal of Microbiological Methods*. 36, 139-145.

[50] Qian, C; Wang, R; Cheng, L; Wang, J. (2010). Theory of microbial carbonate precipitation and its application in restoration of cement-based materials defects. *Chinese Journal of Chemistry*. 28, 847-857.

[51] Kumar B. G, J; Prabhakara, R; Pushpa, H. (2013). Bio mineralisation of calcium carbonate by different bacterial strains and their application in concrete crack remediation. *International Journal of Advances in Engineering and Technology*.

[52] Sierra-Beltran, M. G; Jonkers, H. M; Schlangen, E. (2014). Characterization of sustainable bio-based mortar for concrete repair. *Construction and Building Materials*. 67, 344-352.

[53] Jonkers, H. M; Thijssen, A; Muyzer, G; Copuroglu, O; Schlangen, E. (2010). Application of bacteria as self-healing agent for the development of sustainable concrete. *Ecological Engineering*. 36, 230-235.

[54] Dinku, A; Reinhardt, H. W. (1997). Gas permeability coefficient of cover concrete as a performance control. *Materials and Structures/Materiaux et Constructions*. 30, 387-393.

[55] Achal, V; Mukherjee, A; Reddy, M. S. (2011). Microbial concrete: Way to enhance the durability of building structures. *Journal of Materials in Civil Engineering*. 23 730-734.

[56] Bang, S. S; Galinat, J. K; Ramakrishnan, V. (2001). Calcite precipitation induced by polyurethane-immobilized Bacillus pasteurii. *Enzyme and Microbial Technology*. 28, 404-409.

[57] ASTM C293/C293M-10. (2010). Standard Test Method for Flexural Strength of Concrete (Using Simple Beam With Center-Point Loading). ASTM International, West Conshohocken, PA.

[58] ASTM C78/C78M-15a. (2015). Standard Test Method for Flexural Strength of Concrete (Using Simple Beam with Third-Point Loading), ASTM International, West Conshohocken, PA.

[59] Zhang, M. H; Li, H. (2011). Pore structure and chloride permeability of concrete containing nano-particles for pavement. *Construction and Building Materials*. 25, 608-616.

[60] Basheer, L; Kropp, J; Cleland, D. J. (2001). Assessment of the durability of concrete from its permeation properties: A review. Construction and *Building Materials*. 15, 93-103.

[61] Abbas, A; Carcasses, M.; Ollivier, J. P. (1999). Gas permeability of concrete in relation to its degree of saturation. *Materials and Structures/Materiaux et Constructions*. 32, 3-8.

[62] Basheer, L; Cleland, D. J. (2011). Durability and water absorption properties of surface treated concretes. *Materials and Structures/Materiaux et Constructions*. 44, 957-967.

[63] Sabir, B. B; Wild, S; O'Farrell, M. (1998). A water sorptivity test for mortar and concrete. *Materials and Structures/Materiaux et Constructions*. 31, 568-574.

[64] Tayeh, B. A; Abu Bakar, B. H; Megat Johari, M. A; Yen Lei Voo. (2012). Mechanical and permeability properties of the interface between normal concrete substrate and ultra high performance fiber concrete overlay. *Construction and Building Materials*. 36, 538-548.

[65] Picandet, V; Rangeard, D; Perrot, A; Lecompte, T. (2011). Permeability measurement of fresh cement paste. *Cement and Concrete Research*. 41, 330-338.

[66] Wang, J. Y; Snoeck, D;Van Vlierberghe, S; Verstraete, W; De Belie, N. (2014). Application of hydrogel encapsulated carbonate precipitating bacteria for approaching a realistic self-healing in concrete. *Construction and Building Materials*. 68, 110-119.

[67] ASTM C1585-13. (2013). Standard Test Method for Measurement of Rate of Absorption of Water by Hydraulic-Cement Concretes, ASTM International, West Conshohocken, PA.

[68] Whiting, D. (1981). Rapid determination of the chloride permeability of concrete. Final Report Portland Cement Association, Skokie, IL. Construction Technology Labs.

[69] STM C1202-12. (2012). Standard Test Method for Electrical Indication of Concrete's Ability to Resist Chloride Ion Penetration, ASTM International, West Conshohocken, PA.

[70] Diamond, S. (2000). Mercury porosimetry. An inappropriate method for the measurement of pore size distributions in cement-based materials. *Cement and Concrete Research*. 30, 1517-1525.

[71] Cook, R. A; Hover, K. C. (1991). Mercury porosimetry of hardened cement pastes. *Cement and Concrete Research*. 29, 933-943.

[72] Poon, C. S; Lam, L; Kou, S. C; Wong, Y. L; Wong, R. (2001). Rate of pozzolanic reaction of metakaolin in high-performance cement pastes. *Cement and Concrete Research.* 31, 1301-1306.

[73] Meddah, M. S; Tagnit-Hamou, A. (2009). Pore structure of concrete with mineral admixtures and its effect on self-desiccation shrinkage. *ACI Materials Journal.* 106, 241-250.

[74] Winslow, D. N; Diamond, S. (1970). Mercury prosimetry study of the evolution of porosity in portland cement. *J. Mater.* 5, 564-585.

[75] Wang, L; Sondi, I; Matijević, E. (1999). Preparation of uniform needle-like aragonite particles by homogeneous precipitation. *Journal of Colloid and Interface Science.* 218, 545-553.

[76] Mori, Y; Enomae, T; Isogai, A. (2009). Preparation of pure vaterite by simple mechanical mixing of two aqueous salt solutions. *Materials Science and Engineering C.* 29, 1409-1414.

[77] Dickinson, S. R; McGrath, K. M. (2001). Quantitative determination of binary and tertiary calcium carbonate mixtures using powder X-ray diffraction. *The Analyst.* 126, 1118-1121.

[78] Coats, A. W; Redfern, J. P. (1963). Thermogravimetric analysis. A review. *The Analyst.* 88, 906-924.

[79] Wang, J. Y; Van Tittelboom, K; De Belie, N; Verstraete, W. (2010). Potential of applying bacteria to heal cracks in concrete. in 2nd International Conference on Sustainable Construction Materials and Technologies.

[80] Stuckrath, C; Serpell, R; Valenzuela, L. M; Lopez, M. (2014). Quantification of chemical and biological calcium carbonate precipitation: Performance of self-healing in reinforced mortar containing chemical admixtures. *Cement and Concrete Composites.* 50, 10-15.

[81] Wang, J; Van Tittelboom, K; De Belie, N; Verstraete, W. (2012). Use of silica gel or polyurethane immobilized bacteria for self-healing concrete. *Construction and Building Materials.* 26, 532-540.

[82] Kim, H. S; Park, B. H; Choi, J. H; Yoon, J. S. (2008). Mechanical properties and thermal stability of poly(L-lactide)/calcium carbonate composites. *Journal of Applied Polymer Science.* 109, 3087-3092.

[83] Wang, J; Dewanckele, J; Cnudde, V; Van Vlierberghe, S; Verstraete, W; De Belie, N. (2014). X-ray computed tomography proof of bacterial-based self-healing in concrete. *Cement and Concrete Composites.* 53, 289-304.

[84] Wiktor, V; Jonkers, H. M. (2011). Quantification of crack-healing in novel bacteria-based self-healing concrete. *Cement and Concrete Composites.* 33, 763-770.

[85] Luo, M; Qian, C. X; Li, R. Y. (2015). Factors affecting crack repairing capacity of bacteria-based self-healing concrete. *Construction and Building Materials.* 87, 1-7.

[86] Grattan-Bellew, P. E. (2001). Petrographic and technological methods for evaluation of concrete aggregates, in Handbook of Analytical Techniques in Concrete Science and Technology, Ramachandran, V. S. and Beaudoin, J. J. Editors. William Andrew Publishing: Norwich, NY.

[87] Dick, J; De Windt, W; De Graef, B; Saveyn, H; Van Der Meeren, P; De Belie, N. Verstraete, W. (2006). Bio-deposition of a calcium carbonate layer on degraded limestone by Bacillus species. *Biodegradation.* 17, 357-367.

[88] Kim, H. K; Park, S. J; Han, J. I; Lee, H. K. (2013). Microbially mediated calcium carbonate precipitation on normal and lightweight concrete. *Construction and Building Materials*. 38, 1073-1082.

[89] Vagenas, N. V; Gatsouli, A; Kontoyannis, C. G. (2003). Quantitative analysis of synthetic calcium carbonate polymorphs using FT-IR spectroscopy. *Talanta*. 59, 831-836.

[90] Kontoyannis, C. G; N. V. Vagenas (2000). Calcium carbonate phase analysis using XRD and FT-Raman spectroscopy. *Analyst.* 125, 251-255.

[91] Dandeu, A; Humbert, B; Carteret, C; Muhr, H; Plasari, E; Bossoutrot, J. M. (2006). Raman spectroscopy - A powerful tool for the quantitative determination of the composition of polymorph mixtures: Application to CaCO3 polymorph mixtures. *Chemical Engineering and Technology*. 29, 221-225.

[92] Corbeil, M. C. (2004). Applications of x-ray diffraction in conservation science and archaeometry, International Centre for Diffraction Data. Advances in X-ray Analysis, Volume 47.

[93] De Muynck, W; De Belie, N; Verstraete, W. (2010). Microbial carbonate precipitation in construction materials: A review. *Ecological Engineering*. 36, 118-136.

[94] Achal, V; Mukherjee, A; Basu, P. C; Reddy, M. S. (2009). Lactose mother liquor as an alternative nutrient source for microbial concrete production by Sporosarcina pasteurii. *Journal of Industrial Microbiology and Biotechnology*. 36, 433-438.

[95] Achal, V; Mukherjee, A; Reddy, M. S. (2011). Effect of calcifying bacteria on permeation properties of concrete structures. *Journal of Industrial Microbiology and Biotechnology*. 38, 1229-1234.

In: High Value Processing Technologies
Editors: A. Berenjian, H. Jafarizadeh-Malmiri et al.
ISBN: 978-1-63484-453-6

Chapter 7

PREPARATION AND APPLICATION OF MULTI-ENZYME COMPLEXES AND PECTIC OLIGOSACCHARIDES FROM ORANGE PEEL WASTES

***Pei-jun Li*[1], *Jin-lan Xia*[1,*], *Yang Shan*[2] *and Zhen-yuan Nie*[1]**
[1]School of Minerals Processing and Bioengineering,
Central South University, Changsha, China
[2]Hunan Agriculture Product Processing Institute,
Hunan Academy of Agricultural Sciences, Changsha, China

ABSTRACT

Orange peel wastes (OPW) from the orange processing industry are produced in large quantities worldwide. Many value-added products (e.g., pectin, essential oils, and flavonoids) are extracted from OPW, but the majority of the lignocellulose is currently unutilised and discarded. Recent researches have demonstrated that this lignocellulose could be served as a feedstock for multi-enzyme complexes (MEC) and pectic oligosaccharides (POS) production. The developments of OPW are reviewed in this article. They include production and application of fungal MEC from agricultural wastes, pretreatment techniques enhancing MEC yields and reducing industrial enzyme production cost. In addition, production, refining, fractionation, characterization, and prebiotic activity of POS from OPW are also discussed. This work aims to find a more profitable alternative for the application of orange peel wastes.

Keywords: orange peel waste, multi-enzyme complexes, pectic oligosaccharides, biorefinery, membrane separation

[*] Corresponding author: School of Minerals Processing and Bioengineering, Central South University, Changsha 410083, China, Email: jlxia@csu.edu.cn.

INTRODUCTION

Citrus is one of the most abundant crops in the world. It was reported 88.7 million metric tons of citrus were produced in 2014/15 in the world. It is noted that a great deal of citrus (mainly consists of orange, tangerine/mandarin, grapefruit, lemon/lime, etc.) was processed for production of citrus fruits juice and leaving tremendous orange peel residues, which may constitute about 50% of fresh fruit weight. If the orange peel residues are not processed further, they become wastes and can give rise to many economic and environmental problems because of their fermentability. Table 1 shows the orange and orange peel wastes (OPW) produced in selected countries [1]. In the last decade, OPW have been reported to be an excellent substrate for several value-added products, such as essential oils, pectin, enzymes, single cell protein, natural antioxidants, ethanol, organic acids, and prebiotic oligosaccharides [2, 3]. Figure 1 shows the schematic representation of the biorefinery of orange peel wastes, in which the most worthy to be noted is the production of the multi-enzyme complexes (MEC) that are produced from fermentation of pectin-rich OPW residues after extraction of essential oils, and the MEC can be further used for degrading the solid OPW residues for production of monosaccharides, pectic oligosaccharides as well as dietary fiber, and the former two products are useful for production of bioethanol, antimicrobial or prebiotics materials and agents.

Multi-enzyme complexes (MEC) could be produced using agro-industrial wastes substrates such as wheat bran, orange peel, sugarcane bagasse, and rice straw. Dry citrus peels are rich in pectin, cellulose and hemicellulose, which make it more useful for production of MEC with pectinase, cellulase and xylanase activities, and thus being more advantageous in preparation of pectic oligosaccharides from OPW [4-6].

Pectic oligosaccharides (POS) have been identified as promising prebiotic candidates [7], which have significant effects on modifying both the composition and/or activity of the gastrointestinal microbiota that confers benefits upon host well-being and health [8]. POS can be obtained by depolymerization of OPW by partial enzymatic hydrolysis [9], acid hydrolysis [10], hydrothermal processing [11], and dynamic high-pressure microfluidization [12]. The liquid media from POS production may contain contaminants that should be removed before utilized as food ingredients. The combination of enzymatic depolymerization of various natural polysaccharides and membrane separation was proved to be cost-effective for obtaining prebiotic compounds [13]. After hydrolysis of OPW, fractionation and analysis of POS are usually carried out by chromatographic operations and spectroscopic methods including high-performance size exclusion chromatography (HPSEC), high performance anion exchange chromatography (HPAEC), Fourier transform infrared (FTIR) spectroscopy, and mass spectroscopy [14, 15].

This work provided an updated overview concerning MEC and POS production from orange peel wastes. Moreover, assessments of the prebiotic activities of oligosaccharides were also discussed. It provides more cost-effective ways to produce value-added biotechnological products from OPW.

Table 1. Orange and orange peel wastes (OPW) produced in selected country (1,000 metric Tons) [1]

Country	Orange fruits	Orange processing rates (%)	Estimated OPW
Brazil	16,320	70	5,692
China	6,900	9	325
European Union	6,210	21	657
United States	6,097	71	2,156
Mexico	4,300	30	650

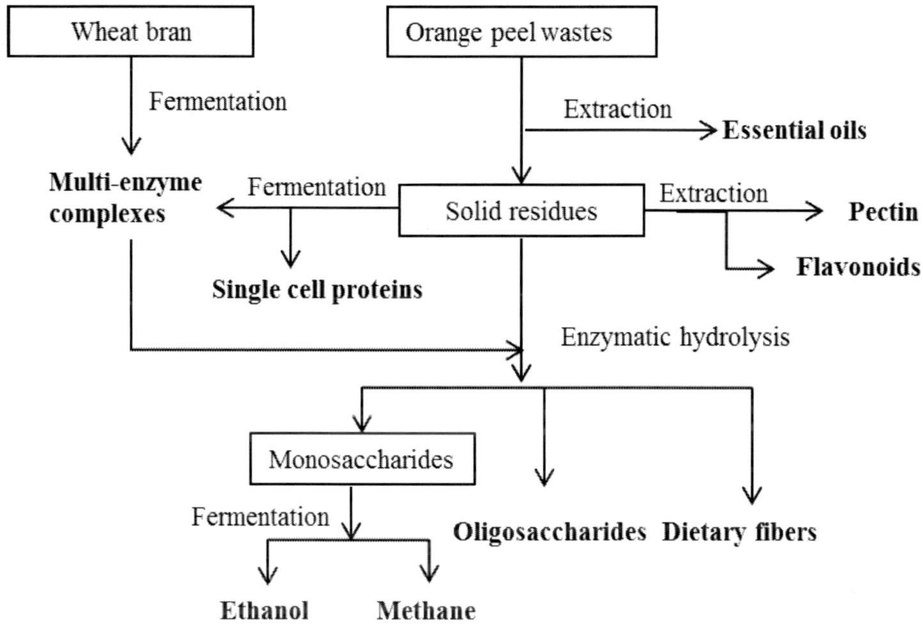

Figure 1. Schematic representation of the biorefinery of orange peel wastes.

Compositional Analyses of OPW

In order to comprehensively harness the waste peel, it is essential to obtain reliable information regarding its chemical composition. Marín et al. [16] reported that the OPW contains cellulose, hemicellulose, pectin, protein, lignin, and ash (Table 2). It is worthy of note that plant polysaccharides (cellulose, hemicellulose, and pectin) occupy about 71 wt% of the OPW, and these plant polysaccharides are suitable for industrial enzymes and oligosaccharides production. The water-insoluble material is mainly composed of glucan and galacturonan. Beside these two polysaccharides, arabinan and galactan are also presented in OPW [9].

MULTI-ENZYME COMPLEXES (MEC) PRODUCTION

Cellulose, hemicellulose, and pectin are the main components of plant cell wall polysaccharides which are widely exist in agricultural residues such as orange peel, wheat bran, orange bagasse, sugarcane bagasse, and rice straw [17]. These low-cost agricultural residues have great potentials as substrates for industrial MEC production [18, 19]. OPW are the most suitable candidate for production of MEC with pectinase and/or polygalaturonase activities because they contain more pectin besides the fermentable sugars (Table 3). MEC production was usually carried out in solid-state (SSF) and submerged fermentation (SmF). Li et al. [20] compared the typical agricultural residues for production of MEC (pectinase, CMCase, and xylanase) by *Aspergillus japonicus* PJ01 in SmF and SSF, and results showed that orange peel and wheat bran were the best substrates for MEC production in SmF and SSF, respectively. Compared to SmF, SSF presents unparalleled advantages, such as higher productivity per reactor volume, lower capital and operating costs, lower space requirements, simpler equipment and easier downstream processing [21]. However, there were still some obstacles needed to overcome, such as unsatisfactory reproducibility of the results, difficulty in scale-up, regulating and monitoring of biomass concentration, and complicated methods of product purification [22].

Production of monomer sugars from orange or banana peels using commercial pectinase, cellulase, and beta-glucosidase has been reported by some teams [41-43]. However, separation and purification of industrial enzymes after the hydrolytic process are expensive, and the optimization of concentration of those enzymes is complicated. MEC have attracted great attentions in biofuel technology as they offered a solution to the effective degradation of complex plant material into fermentable sugars [44]. Recently, the fungi such as *Scytalidium* [45], *Trichoderma* [46], *Aspergillus* [47], and *Penicillium* [48] have been proved effective for MEC production by agro residues fermentation. Fungal MEC have been proved to be effective to hydrolyze domestic waste residues [49], sugarcane bagasse [50], potato peel [51], potato pulp [52], mustard stalk and straw [53]. Up to date, however, there is hardly study about enzymatic hydrolysis of pectin-rich by-products using fungi cocktail enzymes.

Many ways may be used to enhance industrial enzyme yield and reduce enzyme production cost, such as fungi co-cultivation [54], optimization of fermentation conditions [55], and appropriate fermentation mode [24]. In addition to these traditional methods, many reports have shown the stimulative effects of surfactants (nonionic and anionic) on the production and release of enzyme, such as cellulase [56], amylase [57], and xylanase [58]. Tween-80 was the most commonly methodin enhancing the production of cellulases and hemicellulases [59, 58]. Cellulase, pectinase, and xylanase produced by fungi were extracellular enzymes. It was believed that surfactants altered the structure of cell membranes and facilitate extracellular enzyme release [60].

Table 2. Composition of waste orange peel (20% dry basis) [16]

Ash	Sugar	Fat	Protein	Flavonoid	Pectin	Lignin	Cellulose	Hemicellulose
2.56	9.57	4.00	9.06	4.50	23.02	7.52	37.08	11.04

Table 3. Multi-enzyme complexes (MEC) production using orange peel waste as the main substrate

Microorganisms	Substrate[a]	Enzyme	Fermentation[b]	Reference
Aspergillus spp.	WB and OP	Pectinolytic enzymes	SSF	[23]
A. niger	WB and OP	Pectinase and cellulase	SSF and SmF	[24]
A. giganteus	Orange waste, pectin	Polygalacturonase	SmF	[18]
A. awamori	OP and Grape pomace	Exo-polygalacturonase and xylanase	SSF	[5]
A. japonicus	OP	Pectinase, CMCase, FPase, etc.	SmF	[25]
A. foetidus	OP	Pectinase	SmF	[26]
A. fumigatus	OP	Pectinase and polygalacturonase	SSF	[27]
A. sojae	OP	Polygalacturonase	SmF	[28]
Penicillium oxalicum	OP	Pectinase, CMCase, and xylanase	SmF	[29]
P. oxalicum	OP	Pectin lyase	SSF	[30]
P. viridicatum	OP,WB, etc.	Pectin lyase and polygalacturonase	SSF	[31]
Aspergillus and *Penicillium*	OP	Polygalacturonase, endoglucanase	SSF	[32]
Aspergillus and *Penicillium*	Lemon peel pomace	Pectinase	SSF	[33]
A. niger, F. oxysporum,etc.	OP	Cellulase, pectinase and xylanase	SSF	[4]
Trichoderma sp.	OP, etc.	Polygalacturonase, xylanase and lanase	SSF	[34]
Eupenicillium javanicum	Citrus processing wastes	Cellulase, pectinase, xylanase, etc.	SSF	[35]
Pseudozyma sp.	Citrus peel	Pectinase	SSF	[6]
Bacillus licheniformis	Pectin, etc.	Polygalacturonase	SmF	[36]
Thermoascus aurantiacus	OP, WB, etc.	Polygalacturonase	SmF	[37]
Curvularia inaequalis	OP	Pectin lyase	SSF	[38]
Rhizopus oryzae	OP	Pectin lyase	SSF	[39]
Fusarium solani	OP	Exo-polygalacturonase	SSF	[40]

[a] OP, orange peel; WB, wheat bran

[b] SSF, solid state fermentation; SmF, submerged fermentation

Physical pretreatment techniques, such as microwave and ultrasound irradiations were demonstrated effective in removing lignin and hemicellulose, reducing cellulose crystallinity, and increasing the porosity of the materials. Zhao et al. [61] believed that the cracks and porous structure helped the mycelia grow inside the biomass and utilize cellulose as carbon source to synthesize cellulase. Woldesenbet et al. [19] reported that microwave pretreatment of wheat bran led to increase in xylanase yield and the water absorbance by *Bacillus halodurans* FNP 135. Li et al. [25] reported surfactant PEG 4000 and microwave irradiation were environment-friendly and cost-effective ways in enhancing the multienzyme production from agricultural waste orange peel.

PRODUCTION, REFINING, AND FRACTIONATION OF POS

Pectic oligosaccharides (POS) can be obtained by depolymerization of suitable raw materials or purified pectins by partial hydrolysis. Many pectin-rich agro-residues such as OPW, apple pomace, sugar beet, and olive pomace could serve as feedstocks for POS production (Table 4).

As can be seen from Table 4, the mainly hydrolysis modes of pectin-rich agro-residues include autohydrolysis, enzymatic hydrolysis, acid hydrolysis, and dynamic high-pressure microfluidization (DHPM). In contrast to other methods, enzymatic hydrolysis is attractive because it does not produce undesirable by-products, less production of monosaccharides, specific in action as well as it does not require special equipments [81]. Since OPW contain a lot of insoluble polysaccharides (pectin, cellulose and hemicellulose), synergistic action of pectinase, cellulase, and xylanase for its depolymerization is required. Production of POS from pectin-rich biomass using commercial pectinase, cellulase, and beta-glucosidase has been reported by some teams [74, 73, 9, 14]. Martínez Sabajanes et al.[9] assessed the effects of the enzyme loadings (polygalacturonases and cellulases) and reaction time on the conversion of polysaccharides into monosaccharides and oligomers using an empirical model, 31.3 kg of oligosaccharides can be obtained from 100 kg of solid substrates. Combo et al.[79] investigated the hydrolysis of polygalacturonic acid with commercial pectinases allowed the release of the monomer and of several oligomers, depending on the hydrolysis conditions. Olano-Martin et al. [77] investigated the continuous production of POS in an ultrafiltration dead-end membrane reactor using high methylated (60% to 66% methylation) and low methylated (8% methylation) pectin. However, high production cost and low production yield had caused the bottleneck for industrial enzymes application [82], thus alternating enzyme production method using cheaper ingredient with higher yield is the main goal of current studies. Chapla et al. [81] obtained xylooligosaccharides (XOS) from corncobs employing indigenously produced xylanase by *Aspergillus foetidus* MTCC 4898.

As the aqueous hydrolysis of plant polysaccharides is not selective, the resulting reaction liquors must be refined to obtain food grade saccharides. For this purpose, membrane processing has been examined in several studies, owing to its ability to cause purification, fractionation, and/or concentration effects [63, 15]. Ultrafiltration can be used to separate oligosaccharides (OS) from higher molecular weight products or to fractionate OS of different DP. On the other hand, nanofiltration can be useful for concentrating liquors and/or for removing undesired low molecular weight compounds, such as monomeric sugars or phenolic compounds [83]. Amicon stirred cell (model 8400, Millipore) equipped with a regenerated cellulose membrane was widely applied. Holck, et al. [75] employed a regenerated cellulose membrane of 3 kDa Molecular Weight Cut Off (MWCO) for this purpose, whereas Gómez et al. [11] purified oligomers obtained from OPW using a 1 kDa MWCO membrane to retain the high MW compounds. Gavlighi et al. [13] divided the enzymatically degraded products into three molecular weight fractions via membrane separation: HAG1 < 2 kDa; 2 kDa < HAG2 < 10 kDa; HAG3 > 10 kDa. In our previous study, liquid samples from the enzymatic hydrolysis treatment were purified by ultrafiltration, following the scheme shown in Figure 2. Three collected POSs (POS1 < 1kDa; 1 kDa < POS2 < 3 kDa; and POS3 > 3 kDa) were obtained after membrane separation. The use of Amicon module has an important drawback: its low capacity to produce sufficient amounts of refined product to carry out extensive in vitro and/or in vivo studies on their biological properties or to commercialize it. Therefore, in both situations, a scale up is needed and a spiral wound module was employed [11].

The most usual oligosaccharide fractionation technologies are ion exchange chromatography and gel filtration (GF) chromatography. Weak anion-exchange resin Amberlite IRA 96 [84] and Amberlite XAD-2 [85] have been used for removing undesired, non-carbohydrate compounds, whereas Gómez et al. [63] employed DEAE-Sepharose to fractionate OGalA mixtures according their charge. Strong anion-exchange column Source 15Q was also used for POS fractionation [80]. Recently, size exclusion chromatography (SEC) such as Bio-Gel [86, 87], Sephadex G-75 [76], and gel filtration [65] were also used to fractionate oligosaccharide.

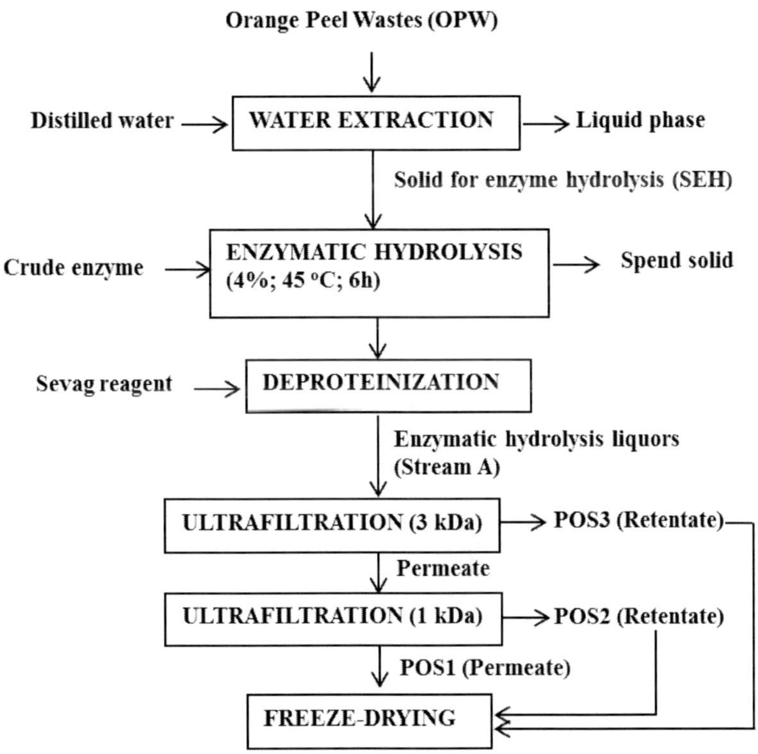

Figure 2. Production and refining of oligosaccharides from orange peel waste.

Table 4. Preparation, fractionation, and characterization of pectic oligosaccharides (POS)

Hydrolysis	Materials	Fractionation	POS characterization[b]	Reference
Autohydrolysis	Orange peel	Membrane and DEAE- Sepharose CL 6B	HPAEC, HPSEC, HILIC-ELSD MS, MALDI-TOF MS	[11]
Autohydrolysis	Orange peel	-	-	[62]
Autohydrolysis	Lemon peel	Membrane filtration	HPAEC	[63]
Autohydrolysis	Olive pomace	Ultrafiltration and Amberlite XAD-16	HPAEC, HPSEC, ESI-MS, MS/MS	[64]
Autohydrolysis	Olive pruning	Gel filtration	HPSEC	[65]
DHPM[a]	Apple pectin	Ultrafiltration	-	[12]
Acid hydrolysis	Orange peel	Nanofiltration	HPAEC	[7]
Enzyme	Bergamot peel	-	HPSEC	[66]
Enzyme	Orange peel	-	HPSEC	[9]
Enzyme	Apple pomace	Membrane filtration	-	[67]
Enzyme	Plant byproducts	-	HPSEC	[68]
Enzyme	Haw fruit	Membrane filtration	-	[69]
Enzyme	Haw pectin	Membrane filtration	-	[70]
Enzyme	Apple pectin	-	TLC	[71]
Enzyme	Sugar beet	-	HILIC–MS,	[72]
Enzyme	Sugar beet	-	HPSEC, TGA/FTIR, XRD	[14]
Enzyme	Sugar beet	-	FTIR	[73]
Enzyme	Sugar beet	-	-	[74]
Enzyme	Pectin	Membrane/Source 15Q	MALDI-TOF/TOF MS	[75]
Enzyme	Ginseng pectin	DEAE–Sepharose, Sephadex G-75	HPGPC, NMR	[76]
Enzyme	Citrus pectin	Membrane reactor	HPSEC	[77]
Enzyme	Lemon pectin	-	3-AQ-HILIC–TWIMMS	[78]
Enzyme	Polygalacturonic acid	-	HPAEC	[79]
Enzyme+alkaline	Potato cell	Source 15Q	HPAEC, HPSEC, NMR	[80]

[a] DHPM, dynamic high-pressure microfluidization

[b] HPAEC, high performance anion exchange chromatography; HPSEC, high-performance size exclusion chromatography; HILIC-ELSD MS, hydrophilic interaction liquid chromatography, evaporative light scattering detector mass spectrometry; MALDI-TOF MS, matrix-assisted laser desorption/ionization-time of flight mass spectrometry; ESI-MS, electrospray ionization mass spectra; TLC, Thin-layer chromatography; TGA/FTIR, thermogravimetric/infrared spectrometry analysis; XRD, X-ray diffraction; FT-IR, Fourier transform infrared spectrophotometer; HPGPC, high-performance gel-permeation chromatography; 3-AQ-HILIC–TWIMMS, 3-aminoquinoline-hydrophilic interaction liquid chromatography-traveling-wave ion mobility mass spectrometry

CHARACTERIZATION OF POS

Recently, chromatographic operations and spectroscopic methods are widely used for characterization of POS. The molecular weight distribution of the target products was assessed by high-performance size exclusion chromatography (HPSEC) [13, 11]. Furthermore, the value of degree of polymerization (DP) could be estimated by HPSEC. In recent years, high-performance anion exchange chromatography with a pulsed amperometric detector (HPAEC-PAD) is recommended for POS analyses [14, 63]. Matrix-assisted laser desorption/ionisation mass spectrometry (MALDI-TOF MS) is an effective tool to provide DP profiles, distribution patterns of the polymers with different molecular weights, and branching information to reveal the whole-level composition of the extracellular polysaccharides [88, 89]. de Oliveira et al. [90] determined the DP values of fructooligosaccharides (FOS) using MALDI-TOF and electrospray ionization mass spectrometry (ESI-MS). Gómez et al. [11] characterized the components of refined POS from OPW by MALDI-TOF MS. Nuclear magnetic resonance (NMR) has been employed in the characterization of rhamnogalacturonan I-type pectic polysaccharides from potato cell wall [80] and ginseng pectin [76]. The composition of POS complex oligomer mixtures has been assessed by HILIC-ELSD MS [11] and 3-AQ-HILIC-TWIMMS [78].

FUNCTIONAL PROPERTIES

The prebiotic effects of pectin and POS depend on the physicochemical characteristics of substrates, and particularly, on the molecular weight distribution and degree of esterification [91]. For instance, Olano-Martin et al.[92] demonstrated that POS with low degree of methylation show higher prebiotic index than high degree of methylation POS. Manderson et al. [7] and Mandalari et al. [8] indicated that pectic oligosaccharides with DP <10 had optimal bifidogenic effects. It is believed that especially butyrate production from oligosaccharides fermentation is beneficial to host health because it is the most important fuel for colonocytes, effects the development and gene expression in intestinal cells, and has a protective role against ulcerative colitis and colon cancer [72].

To assess the prebiotic activity of a given prebiotic candidate, the assimilation kinetics, the distribution of metabolic products and the effects caused on bacterial populations are key aspects to be considered [15]. There are a number of methods to determine the populations of intestinal microorganisms on a test substrate, such as pure culture studies with a range of intestinal bacteria, fermentation vessels containing mixed faecal cultures, and animal models. For a rapid comparative evaluation, batch fermentation vessels inoculated with faecal slurries are useful as they represent the diverse gut microflora [93]. Viability of probiotic bacteria is traditionally assessed by plate counting which has several limitations, including underestimation of cells in aggregates or chains morphology [94]. Alternative methods, such as fluorescent *in situ* hybridization (FISH), nucleic acid amplification techniques such as real-time quantitative PCR (RT-qPCR), reverse transcriptase (RT-PCR) have been used previously in studies of oligosaccharides [95, 96, 11]. Among these methods, RT-qPCR is faster, highly sensitive, specific and enables simultaneous detection of different microorganisms [97].

Conclusion

Orange peel wastes are processing wastes generated from orange juice extraction. This article reviews fungal multi-enzyme complexes (MEC) and pectic oligosaccharides (POS) production from orange peel wastes. MEC have been used for biofuel production as they offered a solution to the effective degradation of complex plant material into fermentable sugars. Pectic oligosaccharides have been identified as promising prebiotic candidates for human health.

Before orange peel lignocellulose can be fully utilised for MEC and POS production, future works have to address certain gaps in the current knowledge. (a) Development of microbial strains which are more efficient at fermenting orange peel into MEC using classic mutagenesis or metabolic engineering. (b) Physical and chemical pretreatment techniques should be assessed for biomass fermentation, with the aim of enhancing the MEC production from agricultural wastes. (c) Specific and highly-efficient MEC should be selected to hydrolysis orange peel into POS, cost-effective separation technologies have to be developed for POS purification. (d) Reliable scientific data information on the interrelationships between POS structure and prebiotic potential is needed.

Acknowledgments

This work was supported by a grant from the National High Technology Research and Development Program of China (863 Program) (No. 2011AA100804), Hunan Provincial Innovation Foundation for Postgraduate (CX2013B084).

References

[1] USDA, (2015). Citrus: World Markets and Trade. Available from http://apps.fas.usda.gov/psdonline/circulars/citrus.pdf.

[2] Ángel Siles López, J., Li, Q. and Thompson, I. P. (2010). Biorefinery of waste orange peel. *Crit. Rev. Biotechnol.*, *30*, 63-69.

[3] Mamma, D. and Christakopoulos, P. (2014). Biotransformation of citrus by-products into value added products. *Waste Biomass Valor.*, *5*, 529-549.

[4] Mamma, D., Kourtoglou, E. and Christakopoulos, P. (2008). Fungal multienzyme production on industrial by-products of the citrus-processing industry. *Bioresour. Technol.*, *99*, 2373-2383.

[5] Díaz, A. B., Bolívar, J., de Ory, I., Caro, I. and Blandino, A. (2011). Applicability of enzymatic extracts obtained by solid state fermentation on grape pomace and orange peels mixtures in must clarification. *LWT Food Sci. Technol.*, *44*, 840-846.

[6] Sharma, S., Mandhan, R. P. and Sharma, J. (2012). Utilization of agro-industrial residues for pectinase production by the novel strain *Pseudozyma* sp. SPJ under solid state cultivation. *Ann. Microbiol.*, *62*, 169-176.

[7] Manderson, K., Pinart, M., Tuohy, K., Grace, W., Hotchkiss, A., Widmer, W., Yadhav, M., Gibson, G. and Rastall, R. (2005). In vitro determination of prebiotic properties of

oligosaccharides derived from an orange juice manufacturing by-product stream. *Appl. Environ. Microbiol.*, *71*, 8383-8389.

[8] Mandalari, G., Palop, C. N., Tuohy, K., Gibson, G., Bennett, R., Waldron, K., Bisignano, G., Narbad, A. and Faulds, C. (2007). *In Vitro* evaluation of the prebiotic activity of a pectic oligosaccharide-rich extract enzymatically derived from bergamot peel. *Appl. Microbiol. Biotechnol.*, *73*, 1173-1179.

[9] Martínez Sabajanes, M., Yáñez, R., Alonso, J. L. and Parajó, J. C. (2012). Pectic oligosaccharides production from orange peel waste by enzymatic hydrolysis. *Int J Food Sci Technol*, *47*, 747-754.

[10] Grohmann, K., Cameron, R. and Buslig, B. (1995). Fractionation and pretreatment of orange peel by dilute acid hydrolysis. *Bioresour. Technol.*, *54*, 129-141.

[11] Gómez, B., Gullón, B., Remoroza, C., Schols, H. A., Parajó, J. C. and Alonso, J. L. (2014). Purification, Characterization, and Prebiotic Properties of Pectic Oligosaccharides from Orange Peel Wastes. *J. Agric. Food Chem.*, *62*, 9769-9782.

[12] Chen, J., Liang, R. h., Liu, W., Li, T., Liu, C. m., Wu, S. s. and Wang, Z. j. (2013). Pectic-oligosaccharides prepared by dynamic high-pressure microfluidization and their *in vitro* fermentation properties. *Carbohydr. Polym.*, *91*, 175-182.

[13] Gavlighi, H. A., Michalak, M., Meyer, A. S. and Mikkelsen, J. D. (2013). Enzymatic depolymerization of gum Tragacanth: Bifidogenic potential of low molecular weight oligosaccharides. *J. Agric. Food Chem.*, *61*, 1272-1278.

[14] Combo, A. M. M., Aguedo, M., Quiévy, N., Danthine, S., Goffin, D., Jacquet, N., Blecker, C., Devaux, J. and Paquot, M. (2013). Characterization of sugar beet pectic-derived oligosaccharides obtained by enzymatic hydrolysis. *Int. J. Biol. Macromol.*, *52*, 148-156.

[15] Gullón, B., Gómez, B., Martínez-Sabajanes, M., Yáñez, R., Parajó, J. and Alonso, J. (2013). Pectic oligosaccharides: Manufacture and functional properties. *Trends Food Sci Technol.*, *30*, 153-161.

[16] Marín, F. R., Soler-Rivas, C., Benavente-García, O., Castillo, J. and Pérez-Alvarez, J. A. (2007). By-products from different citrus processes as a source of customized functional fibres. *Food Chem.*, *100*, 736 741.

[17] Brink, J. and Vries, R. (2011). Fungal enzyme sets for plant polysaccharide degradation. *Appl. Biochem. Biotechnol.*, *91*, 1477-1492.

[18] Pedrolli, D. B., Gomes, E., Monti, R. and Carmona, E. C. (2008). Studies on productivity and characterization of polygalacturonase from *Aspergillus giganteus* submerged culture using citrus pectin and orange waste. *Appl. Biochem. Biotechnol.*, *144*, 191-200.

[19] Woldesenbet, F., Virk, A. P., Gupta, N. and Sharma, P. (2012). Effect of microwave irradiation on xylanase production from wheat bran and biobleaching of eucalyptus kraft pulp. *Appl. Biochem. Biotechnol.*, *167*, 100-108.

[20] Li, P. j., Xia, J. l., Shan, Y. and Nie, Z. y. (2015). Comparative study of multi-enzyme production from typical agro-industrial residues and ultrasound-assisted extraction of crude enzyme in fermentation with *Aspergillus japonicus* PJ01. *Bioprocess Biosyst Eng.*, *38*, 2013-2022.

[21] Pandey, A., Soccol, C. R., Nigam, P. and Soccol, V. T. (2000). Biotechnological potential of agro-industrial residues. I: sugarcane bagasse. *Bioresour. Technol.*, *74*, 69-80.

[22] Hölker, U. and Lenz, J. (2005). Solid-state fermentation—are there any biotechnological advantages? *Curr Opi Microbiol.*, *8*, 301-306.

[23] Heerd, D., Yegin, S., Tari, C. and Fernandez-Lahore, M. (2012). Pectinase enzyme-complex production by *Aspergillus* spp. in solid-state fermentation: A comparative study. *Food Bioprod. Process.*, *90*, 102-110.

[24] Kumar, S., Sharma, H. and Sarkar, B. (2011). Effect of substrate and fermentation conditions on pectinase and cellulase production by *Aspergillus niger* NCIM 548 in submerged (SmF) and solid state fermentation (SSF). *Food Sci. Biotechnol.*, *20*, 1289-1298.

[25] Li, P. j., Xia, J. l., Shan, Y., Nie, Z. y. and Wang, F. r. (2015). Effects of Surfactants and Microwave-assisted Pretreatment of Orange Peel on Extracellular Enzymes Production by *Aspergillus japonicus* PJ01. *Appl. Biochem. Biotechnol.*, *176*, 758-771.

[26] Kumar, S., Jain, N. K., Sharma, K. C., Paswan, R., Mishra, B. K., Srinivasan, R. and Mandhania, S. (2015). Optimization, purification and characterization of pectinases from pectinolytic strain, *Aspergillus foetidus* MTCC 10559. *J Environ Biol.*, *36*, 483-489.

[27] Phutela, U., Dhuna, V., Sandhu, S. and Chadha, B. S. (2005). Pectinase and polygalacturonase production by a thermophilic Aspergillus fumigatus isolated from decomposting orange peels. *Braz. J. Microbiol.*, *36*, 63-69.

[28] Gogus, N., Taze, B. H., Demir, H., Tari, C., Unluturk, S. and Lahore, M. F. (2014). Evaluation of orange peel, an industrial waste, for the production of *Aspergillus sojae* polygalacturonase considering both morphology and rheology effects. *Turk. J. Biol.*, *38*, 537-548.

[29] Li, P. J., Xia, J. L., Shan, Y., Nie, Z. Y., Su, D. L., Gao, Q. R., Zhang, C. and Ma, Y. L. (2015). Optimizing Production of Pectinase from Orange Peel by *Penicillium oxalicum* PJ02 Using Response Surface Methodology. *Waste Biomass Valor.*, *6*, 13-22.

[30] Yadav, S. and Shastri, N. V. (2007). Purification and properties of an extracellular pectin lyase produced by the strain of *Penicillium oxalicum* in solid-state fermentation. *Indian J. Biochem. Biophys.*, *44*, 247-251.

[31] Silva, D., Martins, E. S., da Silva, R. and Gomes, E. (2002). Pectinase production by Penicillium viridicatum RFC3 by solid state fermentation using agricultural wastes and agro-industrial by-products. *Braz. J. Microbiol.*, *33*, 318-324.

[32] Adeleke, A. J., Odunfa, S. A., Olanbiwonninu, A. and Owoseni, M. C. (2012). Production of Cellulase and Pectinase from Orange Peels by Fungi. *Nature and Science*, *10*, 107-112.

[33] Ruiz, H. A., Rodríguez-Jasso, R. M., Rodríguez, R., Contreras-Esquivel, J. C. and Aguilar, C. N. (2012). Pectinase production from lemon peel pomace as support and carbon source in solid-state fermentation column-tray bioreactor. *Biochem. Eng. J.*, *65*, 90-95.

[34] Mohamed, S. A., Al-Malki, A. L., Khan, J. A., Kabli, S. A. and Al-Garni, S. M. (2013). Solid State Production of Polygalacturonase and Xylanase by *Trichoderma* Species Using Cantaloupe and Watermelon Rinds. *J. Microbiol.*, *51*, 605-611.

[35] Tao, N. g., Shi, W. q., Liu, Y. j. and Huang, S. r. (2011). Production of feed enzymes from citrus processing waste by solid-state fermentation with *Eupenicillium javanicum*. *Int J Food Sci Technol.*, *46*, 1073-1079.

[36] Rehman, H. U., Qader, S. A. U. and Aman, A. (2012). Polygalacturonase: Production of pectin depolymerising enzyme from *Bacillus licheniformis* KIBGE IB-21. *Carbohydr. Polym.*, *90*, 387-391.

[37] Martins, E. d. S., Leite, R. S. R., da Silva, R. and Gomes, E. (2012). Production and characterization of polygalacturonase from thermophilic *Thermoascus aurantiacus* on submerged fermentation. *Ann. Microbiol.*, 1-7.

[38] Afifi, A. F., Fawzi, E. M. and Foaad, M. A. (2002). Purification and characterization of a pectin lyase produced by *Curvularia inaequalis* NRRL 13884 on orange peels waste, solid state culture. *Ann. Microbiol.*, *52*, 287-297.

[39] Hamdy, H. S. (2005). Purification and characterization of the pectin lyase produced by *Rhizopus oryzae* grown on orange peels. *Ann. Microbiol.*, *55*, 205-211.

[40] Hamid, S., Bhatti, H. N. and Qayyum, U. A. (2008). Enhanced production of exo-polygalacturonase by *Fusarium solani* in solid-state fermentation. *Asian J Chem.*, *20*, 4273-4281.

[41] Wilkins, M. R., Widmer, W. W., Grohmann, K. and Cameron, R. G. (2007). Hydrolysis of grapefruit peel waste with cellulase and pectinase enzymes. *Bioresour. Technol.*, *98*, 1596-1601.

[42] Ghorbanpour Khamseh, A. A. and Miccio, M. (2012). Comparison of batch, fed-batch and continuous well-mixed reactors for enzymatic hydrolysis of orange peel wastes. *Process Biochem.*, *47*, 1588-1594.

[43] Oberoi, H. S., Sandhu, S. K. and Vadlani, P. V. (2012). Statistical optimization of hydrolysis process for banana peels using cellulolytic and pectinolytic enzymes. *Food Bioprod. Process.*, *90*, 257-265.

[44] van Dyk, J. S., Sakka, M., Sakka, K. and Pletschke, B. I. (2010). Characterisation of the multi-enzyme complex xylanase activity from *Bacillus licheniformis* SVD1. *Enzyme Microb. Technol.*, *47*, 174-177.

[45] Jatinder, K., Chadha, B. and Saini, H. (2006). Optimization of culture conditions for production of cellulases and xylanases by *Scytalidium thermophilum* using response surface methodology. *World J Microbiol Biotechnol.*, *22*, 169-176.

[46] Sukumaran, R. K., Singhania, R. R., Mathew, G. M. and Pandey, A. (2009). Cellulase production using biomass feed stock and its application in lignocellulose saccharification for bio-ethanol production. *Renew Energy.*, *34*, 421-424.

[47] López, J. A., Lázaro, C. d. C., Castilho, L. d. R., Freire, D. M. G. and Castro, A. M. d. (2013). Characterization of multienzyme solutions produced by solid-state fermentation of babassu cake, for use in cold hydrolysis of raw biomass. *Biochem. Eng. J.*, *77*, 231-239.

[48] Gonçalves, D., Teixeira, J. A., Bazzolli, D. S., de Queiroz, M. and de Araújo, E. (2012). Use of response surface methodology to optimize production of pectinases by recombinant *Penicillium griseoroseum* T20. *Biocatal Agric Biotechnol.*, *1*, 140-146.

[49] Bansal, N., Tewari, R., Gupta, J. K., Soni, R. and Soni, S. K. (2011). A novel strain of *Aspergillus niger* producing a cocktail of hydrolytic depolymerising enzymes for the production of second generation biofuels. *BioResources.*, *6*, 552-569.

[50] Gottschalk, L. M. F., Oliveira, R. A. and Bon, E. P. d. S. (2010). Cellulases, xylanases, β-glucosidase and ferulic acid esterase produced by *Trichoderma* and *Aspergillus* act synergistically in the hydrolysis of sugarcane bagasse. *Biochem. Eng. J.*, *51*, 72-78.

[51] Khawla, B. J., Sameh, M., Imen, G., Donyes, F., Dhouha, G., Raoudha, E. G. and Oumèma, N. E. (2014). Potato peel as feedstock for bioethanol production: A comparison of acidic and enzymatic hydrolysis. *Ind Crops Prod.*, *52*, 144-149.

[52] Gao, M. T., Yano, S. and Minowa, T. (2014). Characteristics of enzymes from *Acremonium cellulolyticus* strains and their utilization in the saccharification of potato pulp. *Biochem. Eng. J.*, *83*, 1-7.

[53] Pal, S., Banik, S. P. and Khowala, S. (2013). Mustard stalk and straw: A new source for production of lignocellulolytic enzymes by the fungus *Termitomyces clypeatus* and as a substrate for saccharification. *Ind Crops Prod.*, *41*, 283-288.

[54] Hu, H. L., van den Brink, J., Gruben, B. S., Wösten, H. A. B., Gu, J. D. and de Vries, R. P. (2011). Improved enzyme production by co-cultivation of *Aspergillus niger* and *Aspergillus oryzae* and with other fungi. *Int. Biodeterior. Biodegradation*, *65*, 248-252.

[55] Sharma, D. and Satyanarayana, T. (2006). A marked enhancement in the production of a highly alkaline and thermostable pectinase by *Bacillus pumilus* dcsr1 in submerged fermentation by using statistical methods. *Bioresour. Technol.*, *97*, 727-733.

[56] Pardo, A. G. (1996). Effect of surfactants on cellulase production by *Nectria catalinensis*. *Curr Microbiol.*, *33*, 275-278.

[57] Reddy, R. M., Reddy, P. G. and Seenayya, G. (1999). Enhanced production of thermostable β-amylase and pullulanase in the presence of surfactants by *Clostridium thermosulfurogenes* SV2. *Process Biochem.*, *34*, 87-92.

[58] Zeng, G. M., Shi, J. G., Yuan, X. Z., Liu, J., Zhang, Z. B., Huang, G. H., Li, J. B., Xi, B. D. and Liu, H. L. (2006). Effects of Tween 80 and rhamnolipid on the extracellular enzymes of *Penicillium simplicissimum* isolated from compost. *Enzyme Microb. Technol.*, *39*, 1451-1456.

[59] Tangnu, S. K., Blanch, H. W. and Wilke, C. R. (1981). Enhanced production of cellulase, hemicellulase, and β-glucosidase by *Trichoderma reesei* (Rut C-30). *Biotechnol. Bioeng.*, *23*, 1837-1849.

[60] Ahuja, S., Ferreira, G. and Moreira, A. (2004). Production of an endoglucanase by the shipworm bacterium, *Teredinobacter turnirae*. *J. Ind. Microbiol. Biotechnol.*, *31*, 41-47.

[61] Zhao, X., Zhou, Y., Zheng, G. and Liu, D. (2010). Microwave pretreatment of substrates for cellulase production by solid-state fermentation. *Appl. Biochem. Biotechnol.*, *160*, 1557-1571.

[62] Martínez, M., Yáñez, R., Alonsó, J. L. and Parajó, J. C. (2010). Chemical production of pectic oligosaccharides from orange peel wastes. *Ind Eng Chem Res.*, *49*, 8470-8476.

[63] Gómez, B., Gullón, B., Yáñez, R., Parajó, J. C. and Alonso, J. L. (2013). Pectic Oligosacharides from Lemon Peel Wastes: Production, Purification, and Chemical Characterization. *J. Agric. Food Chem.*, *61*, 10043-10053.

[64] Lama-Muñoz, A., Rodríguez-Gutiérrez, G., Rubio-Senent, F. and Fernández-Bolaños, J. (2012). Production, characterization and isolation of neutral and pectic oligosaccharides with low molecular weights from olive by-products thermally treated. *Food Hydrocoll.*, *28*, 92-104.

[65] Cara, C., Ruiz, E., Carvalheiro, F., Moura, P., Ballesteros, I., Castro, E. and Gírio, F. (2012). Production, purification and characterisation of oligosaccharides from olive tree pruning autohydrolysis. *Ind Crops Prod.*, *40*, 225-231.

[66] Mandalari, G., Bennett, R. N., Kirby, A. R., Lo Curto, R. B., Bisignano, G., Waldron, K. W. and Faulds, C. B. (2006). Enzymatic hydrolysis of flavonoids and pectic

oligosaccharides from bergamot (*Citrus bergamia* Risso) peel. *J. Agric. Food Chem.*, *54*, 8307-8313.

[67] Gullón, B., Gullón, P., Sanz, Y., Alonso, J. L. and Parajó, J. C. (2011). Prebiotic potential of a refined product containing pectic oligosaccharides. *LWT Food Sci. Technol.*, *44*, 1687-1696.

[68] Zykwinska, A., Boiffard, M. H., Kontkanen, H., Buchert, J., Thibault, J. F. and Bonnin, E. (2008). Extraction of Green Labeled Pectins and Pectic Oligosaccharides from Plant Byproducts. *J. Agric. Food Chem.*, *56*, 8926-8935.

[69] Li, T. P., Liu, Y. H., Dong, Y. P., Li, S. H. and Zhu, R. G. (2014). Anti-fat deposition and antioxidant effects of haw pectic oligosaccharide in the liver of high-fat-fed mice. *Cyta-Journal of Food.*, *12*, 27-31.

[70] Li, T., Li, S., Du, L., Wang, N., Guo, M., Zhang, J., Yan, F. and Zhang, H. (2010). Effects of haw pectic oligosaccharide on lipid metabolism and oxidative stress in experimental hyperlipidemia mice induced by high-fat diet. *Food Chem.*, *121*, 1010-1013.

[71] Nikolić, M. V. and Mojovic, L. (2007). Hydrolysis of apple pectin by the coordinated activity of pectic enzymes. *Food Chem.*, *101*, 1-9.

[72] Leijdekkers, A. G. M., Aguirre, M., Venema, K., Bosch, G., Gruppen, H. and Schols, H. A. (2014). *In vitro* Fermentability of Sugar Beet Pulp Derived Oligosaccharides Using Human and Pig Fecal Inocula. *J. Agric. Food Chem.*, *62*, 1079-1087.

[73] Concha Olmos, J. and Zúñiga Hansen, M. E. (2012). Enzymatic depolymerization of sugar beet pulp: Production and characterization of pectin and pectic-oligosaccharides as a potential source for functional carbohydrates. *Chem Eng J.*, *192*, 29-36.

[74] Martinez, M., Gullon, B., Yanez, R., Alonso, J. L. and Parajó, J. C. (2009). Direct enzymatic production of oligosaccharide mixtures from sugar beet pulp: experimental evaluation and mathematical modeling. *J. Agric. Food Chem.*, *57*, 5510-5517.

[75] Holck, J., Hjernø, K., Lorentzen, A., Vigsnæs, L. K., Hemmingsen, L., Licht, T. R., Mikkelsen, J. D. and Meyer, A. S. (2011). Tailored enzymatic production of oligosaccharides from sugar beet pectin and evidence of differential effects of a single DP chain length difference on human faecal microbiota composition after *in vitro* fermentation. *Process Biochem.*, *46*, 1039-1049.

[76] Yu, L., Zhang, X., Li, S., Liu, X., Sun, L., Liu, H., Iteku, J., Zhou, Y. and Tai, G. (2010). Rhamnogalacturonan I domains from ginseng pectin. *Carbohydr. Polym.*, *79*, 811-817.

[77] Olano-Martin, E., Mountzouris, K., Gibson, G. and Rastall, R. (2001). Continuous production of pectic oligosaccharides in an enzyme membrane reactor. *J. Food Sci.*, *66*, 966-971.

[78] Leijdekkers, A. G. M., Huang, J. H., Bakx, E. J., Gruppen, H. and Schols, H. A. (2015). Identification of novel isomeric pectic oligosaccharides using hydrophilic interaction chromatography coupled to traveling-wave ion mobility mass spectrometry. *Carbohydr. Res.*, *404*, 1-8.

[79] Combo, A. M. M., Aguedo, M., Goffin, D., Wathelet, B. and Paquot, M. (2012). Enzymatic production of pectic oligosaccharides from polygalacturonic acid with commercial pectinase preparations. *Food Bioprod. Process.*, *90*, 588-596.

[80] Khodaei, N. and Karboune, S. (2013). Extraction and structural characterisation of rhamnogalacturonan I-type pectic polysaccharides from potato cell wall. *Food Chem.*, *139*, 617-623.

[81] Chapla, D., Pandit, P. and Shah, A. (2012). Production of xylooligosaccharides from corncob xylan by fungal xylanase and their utilization by probiotics. *Bioresour. Technol.*, *115*, 215-221.

[82] Kang, S. W., Park, Y. S., Lee, J. S., Hong, S. I. and Kim, S. W. (2004). Production of cellulases and hemicellulases by *Aspergillus niger* KK2 from lignocellulosic biomass. *Bioresour. Technol.*, *91*, 153-156.

[83] Gullón, P., González-Muñoz, M. J., Domínguez, H. and Parajó, J. C. (2008). Membrane processing of liquors from *Eucalyptus globulus* autohydrolysis. *J Food Eng.*, *87*, 257-265.

[84] Gullón, B., Gullón, P., Tavaria, F., Pintado, M., Gomes, A. M., Alonso, J. L. and Parajó, J. C. (2014). Structural features and assessment of prebiotic activity of refined arabinoxylooligosaccharides from wheat bran. *J Funct Foods.*, *6*, 438-449.

[85] Yuan, X., Wang, J. and Yao, H. (2006). Production of feruloyl oligosaccharides from wheat bran insoluble dietary fibre by xylanases from *Bacillus subtilis*. *Food Chem.*, *95*, 484-492.

[86] Hu, T., Li, C., Zhao, X., Li, G., Yu, G. and Guan, H. (2013). Preparation and characterization of guluronic acid oligosaccharides degraded by a rapid microwave irradiation method. *Carbohydr. Res.*, *373*, 53-58.

[87] Chidouh, A., Aouadi, S. and Heyraud, A. (2014). Extraction, fractionation and characterization of water-soluble polysaccharide fractions from myrtle (*Myrtus communis* L.) fruit. *Food Hydrocoll.*, *35*, 733-739.

[88] Harvey, D. J. (2003). Matrix-assisted laser desorption/ionization mass spectrometry of carbohydrates and glycoconjugates. *Int J Mass Spectrom.*, *226*, 1-35.

[89] Liang, J. R., Ai, X. X., Gao, Y. H. and Chen, C. P. (2013). MALDI-TOF MS analysis of the extracellular polysaccharides released by the diatom *Thalassiosira pseudonana*. *J. Appl. Microbiol.*, *25*, 477-484.

[90] de Oliveira, A. J. B., Gonçalves, R. A. C., Chierrito, T. P. C., dos Santos, M. M., de Souza, L. M., Gorin, P. A. J., Sassaki, G. L. and Iacomini, M. (2011). Structure and degree of polymerisation of fructooligosaccharides present in roots and leaves of *Stevia rebaudiana* (Bert.) Bertoni. *Food Chem.*, *129*, 305-311.

[91] Gullón, P., Gullón, B., Moure, A., Alonso, J. L., Domínguez, H. and Parajó, J. C. (2009). Manufacture of prebiotics from biomass sources, Prebiotics and Probiotics Science and Technology. Springer, 535-589.

[92] Olano-Martin, E., Gibson, G. and Rastall, R. (2002). Comparison of the *in vitro* bifidogenic properties of pectins and pectic-oligosaccharides. *J. Appl. Microbiol.*, *93*, 505-511.

[93] Rycroft, C. E., Jones, M. R., Gibson, G. R. and Rastall, R. A. (2001). A comparative *in vitro* evaluation of the fermentation properties of prebiotic oligosaccharides. *J. Appl. Microbiol.*, *91*, 878-887. doi: 10. 1046/j.1365-2672.2001.01446.x.

[94] Reimann, S., Grattepanche, F., Rezzonico, E. and Lacroix, C. (2010). Development of a real-time RT-PCR method for enumeration of viable *Bifidobacterium longum* cells in different morphologies. *Food Microbiol.*, *27*, 236-242.

[95] Holck, J., Lorentzen, A., Vigsnæs, L. K., Licht, T. R., Mikkelsen, J. D. and Meyer, A. S. (2011). Feruloylated and nonferuloylated arabino-oligosaccharides from sugar beet pectin selectively stimulate the growth of *Bifidobacterium* spp. in human fecal in vitro fermentations. *J. Agric. Food Chem.*, *59*, 6511-6519.

[96] Yang, L. C., Lin, W. C. and Lu, T. J. (2012). Characterization and Prebiotic Activity of Aqueous Extract and Indigestible Polysaccharide from *Anoectochilus formosanus*. *J. Agric. Food Chem.*, *60*, 8590-8599.

[97] Postollec, F., Falentin, H., Pavan, S., Combrisson, J. and Sohier, D. (2011). Recent advances in quantitative PCR (qPCR) applications in food microbiology. *Food Microbiol.*, *28*, 848-861.

In: High Value Processing Technologies
Editors: A. Berenjian, H. Jafarizadeh-Malmiri et al.
ISBN: 978-1-63484-453-6

Chapter 8

ISOLATION PROCESS OF ANTIMICROBIAL COMPOUNDS FROM PLANTS

***Gbago Onivogui*[1,2] *and Yuanda Song*[1,3,*]**
[1]Center for Functional Foods and Human Health,
School of Food Science and Technology, Jiangnan University,
Wuxi, Jiangsu, China
[2]Center of Scientific research Rogbane Conakry, Conakry, Guinea
[3]Colin Ratledge Center for Microbial Lipids,
School of Agriculture Engineering and Food Science,
Shandong University of Technology, Zibo, Shandong, China

ABSTRACT

In the last few years, the identification and development of bioactive compounds from different plants has become a major area of health and medical-related research. This chapter provides an updated and comprehensive overview on the process of isolation and identification of antimicrobial compounds from plants. Recently, several studies have shown the identification and isolation process of antifungal and antibacterial compounds from different plant parts. As choosing the extraction methods including the types of extracting solvents play a most important role in the isolation and identification process of bioactive constituents from plants, several chromatography methods for the isolation and identification of antimicrobial compounds including thin layer chromatography, high performance chromatography; gas chromatography-mass analysis, liquid chromatography-mass analysis, and nuclear magnetic resonance analysis have been reviewed in this chapter.

* Corresponding author: Yuanda Song, Colin Ratledge Center for Microbial Lipids, School of Agriculture Engineering and Food Science Shandong University of Technology. Email: ysong@sdut.edu.cn.

INTRODUCTION

Plants have formed the basis of sophisticated traditional medicinal systems that have been in existence for thousands of years, and they continue to provide humanity with new remedies [1, 2]. Medicinal plants have been used in centuries as a remedy for human diseases because they contain components of therapeutic value. Plants produce a diverse range of bioactive molecules, making them rich source of different types of medicines [3]. The continuously increase in resistance of human pathogenic microorganisms to the known antibiotic leads to necessity for searching new sources for production of new active antimicrobial compounds from the wild plants, since many wild plants have been found to be able to produce antimicrobial compounds as a defense phenomenon against invading microorganisms.

The current advancement in science has made it possible for the isolation of compounds of medical importance from plants that are used in traditional medicinal practices. Different parts of plants such as leaves, roots, stems barks and rhizomes are often extracted using different solvents. In most cases, extracts have been shown to be biologically active in both in vitro and in vivo test systems [4]. Despite of substantial developments of extraction and separation techniques, isolation of these bioactive compounds from plants are still a challenging task. Isolation of pure, pharmacologically active constituents from plants remains a long and tedious process. For this reason, it is necessary to have methods available which eliminate unnecessary separation procedures. In the recent research, many scientists focused on determining the antimicrobial activity of plant extracts found in folk medicine, essential oils or isolated compounds such as alkaloids, flavonoids, sesquiterpene lactones, diterpenes and triterpenes [5]. The medicinal chemists have shunned natural products for a few reasons, the process of isolation and identification of new natural products can be costly and is generally time consuming, and as well the structures can be quite complex and deviate from the aforementioned Rule-of-Five. Compared to combinatorially designed compounds natural products tend to have: 1) More single bonds, protonated amino and free hydroxyl groups, and fewer aromatic groups. 2) There is a great diversity of cyclic systems, which are often quite rigid. 3) Often several chiral centers and fewer rotatable double bonds are present [6]. This chapter will focus on process of identification and isolation of antimicrobials compounds from plants. Drug discovery process from medicinal plants, extraction methods, types of solvent for extracting and separation methods are discussed.

DRUG DISCOVERY PROCESS FROM MEDICINAL PLANTS

In pharmaceutical sector, drug discovery process begins with target identification, target prioritization, lead identification and lead optimization. This cycle involves various aspects like preclinical technologies which covers preliminary in vitro and in vivo assays for selection of therapeutically active compounds, chemistry manufacturing and controls (CMC)/ pharmaceutics to formulate the drugs by complete studies of its property and intended for clinical use, pharmacology/toxicology to determine efficacy and safety and clinical investigation of three stages mentioning effect on humans and ultimately approval as NDA (New Drug Application) as well as MAA (Marketing Authorization Application). The drug development process is becoming more complex and capital intensive, and companies remain 'target rich' but 'lead poor,' with lead discovery as a greater bottleneck. It is estimated that, to develop one successful drug, 12-15 years and US $ 900 million are required. Lag phase for botanical medicine is now rapidly changing for a number of reasons. Problems with drug-resistant microorganisms, side effects of modern drugs, and emerging diseases where no medicines are available, have stimulated renewed interest in plants as a significant source of new medicines and the process of drug discovery from plants (Figure 1) begins with a botanist, ethno botanist, ethno pharmacologist, or plant ecologist who collects and identifies the plants of interest. Collection may involve species with known biological activity for which active compound(s) have not been isolated (i.e., traditionally used herbal remedies). It is necessary to respect the intellectual property rights of a given country where plant(s) of interest are collected [7]. Phytochemists (natural product chemists) prepare extracts from the plant material, subject these extracts to biological screening in pharmacologically relevant assays, and commence the process of isolation and characterization of the active compound(s) through bioassay-guided fractionation. Molecular biology has become essential to medicinal plant drug discovery through the determination and implementation of appropriate screening assays directed towards physiologically relevant molecular targets. Pharmacognosy encapsulates all of these fields into a distinct interdisciplinary science.

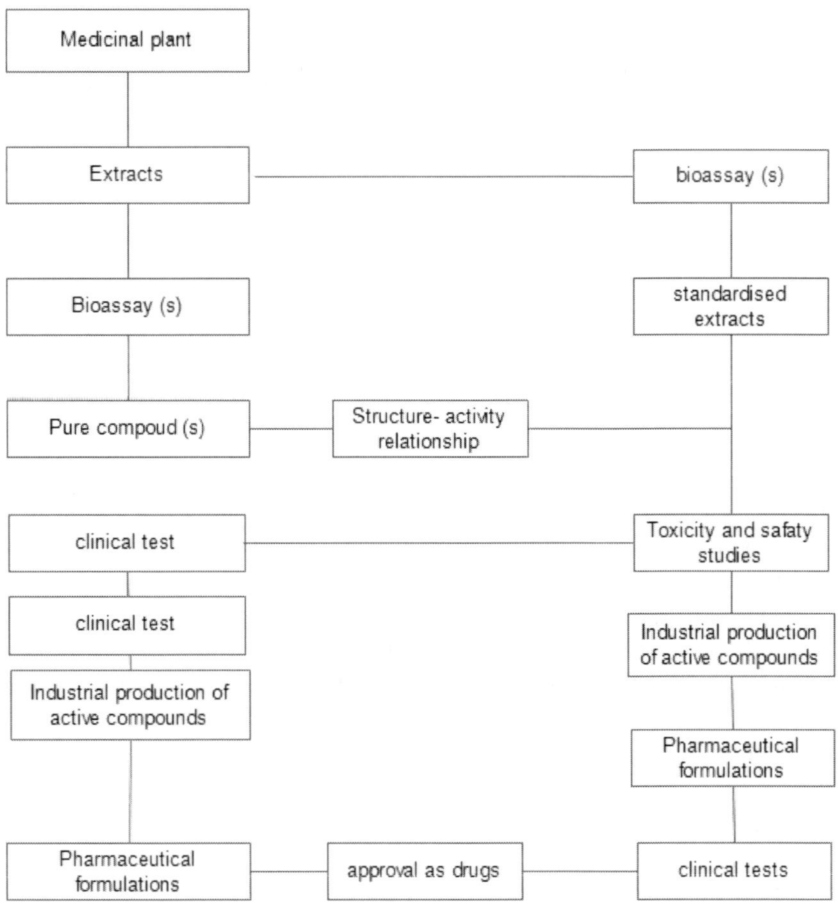

Figure 1. Drug Discovery from Medicinal Plants.

EXTRACTION METHODS

Extraction is the crucial first step in the analysis of medicinal plants, because it is necessary to extract the desired antimicrobial compounds from the plant materials for further separation and characterization (Table 1). Both the solvent and extraction system may affect the final results. The basic operation included steps, such as pre-washing, drying of plant materials or freeze drying, grinding to obtain a homogenous sample and often improving the kinetics of analytic extraction and also increasing the contact of sample surface with the solvent system. Proper actions must be taken to assure that potential active constituents are not lost or destroyed during the preparation of the extract from plant samples. If the plant was selected on the basis of traditional uses [8], the most appropriate method would be that in which the extracts were the same as that used in folk medicine or phytotherapy. The method of monitoring the isolation process is dependent on the compounds of interest. If the extraction was performed in search of a known compound, then the desired compounds should be available to use as standards. When isolating an unknown compound from an extract with observed biological activity, then an activity-guided approach can be used. The appropriate screening method can be selected so only the fractions containing the compound of interest are carried through the entire extraction and isolation process.

Bioactive compounds including antimicrobial agents comprise a very wide range of heterogeneous phytochemicals compounds such as phenolic compounds, alkaloids, aglycones and volatile compounds with different chemical and structural features. They have been extracted from plants by several scientists using different extraction methods and solvents.

Extraction of antimicrobial compounds from medicinal plants and spices can be carried out in a variety of ways. A number of methods have been developed in recent years such as traditional extractions (soxhlet, infusion and maceration), and supercritical fluid extraction. The application of enhanced extraction techniques using non-toxic solvents such as supercritical carbon dioxide extraction (SFE), microwave-assisted extraction (MAE) and pressurized fluid extraction (PFE) or accelerated solvent extraction (ASE) with ethanol–water solvent mixtures were applied in the extraction of phenolic compounds from plant materials and the conventional soxhlet extraction method have been used for extraction of several bioactive compounds.

Supercritical fluid extraction is a well-known method for producing high quality plant extracts in a safe and clean way. On the other hand, MAE extraction has been developed as a simple, faster and less solvent consumption method for the production of high valuable extracts from plant materials. In addition, the use of pressure was introduced in microwave-assisted extraction (P-MAE) to further enhance the extraction of phytochemical compounds from the medicinal plants. Theoretically, increases of pressure and the corresponding increments in extraction temperature could increase the solubility of the bioactive compounds in the extracting solvent.

Ultrasonic-assisted extraction (UAE) has been proven to be one of the most important techniques for extracting bioactive compounds from plants, and it is quite adaptable on a small or large scale, beside is a potential technology due to the cheaper and fewer instrumental requirements. Application of UAE offers many advantages including: fewer amounts of solvent, lower temperatures and short extraction time, which is very useful for extraction of thermo-labile and unstable compounds and also increased extraction yield [9].

Table 1. Some of the experimental condition for several methods of extraction for plants

Plants	Plant parts	Solvents	Methods	Compounds	Action or clinical use Microorganism	Ref.
Trichodesma indicum (Linn.) R. Br.	Root	Ethanol	Maceration	n-decanyl laurate, n-tetradecanyl laurate, n-nonacosanyl palmitate, stigmast-5-en-3b-ol-21(24) olide, n-pentacos-9-one, n-dotriacont-9-one-13-ene, stigmast-5-en-3b-ol-23-one, lanast-5-en-3beta-D-glucopyranosyl-21 (24)-olide	*S. aureus, B. subtilis, C. albicans*	[27]
Guava (Psidium guajava L.)	Leaves	Methanol-water	Maceration	Morin-3-O-alpha-L-lyxopyranoside, morin-3-O-a-L-arabopyranoside, morin, quercetin, quercetin-3-O-arabinoside	*S. enteritidis, B. cereus*	[28]
Clinacanthus nutans Lindau	Leaves Stems	Ethanol Ethanol-water	MEA, *SFE*, Soxhlet	Neophytadiene, myristic acid, palmitic acid acid methyl ester, stearic acid methyl ester, squalene, linoleic acid ethyl ester, linolenic acid methyl ester	*E. coli, S. aureus, B. subtilis*	[15]
Carissa opaca	Roots	Methanolic Ethyl acetate	Maceration cold	limonene, 2' hydroxyacetophenone, vanillin, naphthalenone, 2,3,3-trimethyl-2-(3-methylbuta-1,3-dienyl)-6-methylenecyclo-hexanone, 2-benzenedicarboxylic acid, mono(2-ethylhexyl) ester, b-sitosterol, vitamin E, rutin, quercetin, lupeol, epigallocatechin,	*B. subtilis, E. coli, P. aeruginosa, S. aureus, S. typhi, Enterobacter cloacae*	[29]
F. vulgaris, F vaillantii	Leaves	Methanol-water	Maceration	Carvacrol Fumaric acids	*B cereus, S. marcescens, E. coli, S. Aureus, E. Aerogenes, K. Pneumonia, P. aeruginosa, P. vulgaris, B. megaterium*	[30]
Moringa oleifera	Seeds	Ethanol Methanol Acetone Hexane Chloroform	Maceration	Gallic acid, p-coumaric acid, ferulic acid, caffeic acid, protocatechuic acid, cinnamic acid, catechin, epicatechin. vanillin, quercetin	*S. aureus, B. cereus, E. coli, Y. enterocolitica*	[31]
Blueberry	Leaves	Ethanol Acetone Methanol	Maceration	Gallic acid, protocatechuic, caffeic acid, vanillic acid, syringic acid, p-coumaric, ferulic acids, quercetin, kaempferol	*S. aureus, L. monocytogenes, S. Typhimurium, E. coli, Monilina Vaccinii-corybosi*	[32]

Table 1. (Continued)

Plants	Plant parts	Solvents	Methods	Compounds	Action or clinical use Microorganism	Ref.
Bauhinia purpurea	Leaves	Ethanol	Soxhlet Maceration Ultrasonic	Phenolics, flavonoids, tannins	*S. aureus, B. cereus, B. subtilis, E. coli, P. aeruginosa, K. Pneumoniae, Shigella*	[17]
T. acutifolius	Leaves	Water	Hot water infusion	Rutin, nicotiflorin, hyperoside, isoquercitrin, tripodantoside	*S. aureus, P. aeruginosa*	[33]
Seabuckthorn	*Leaves*	70% ethanol	Maceration	Gallic acid, Rutin, Quercetin-3-galactoside, Quercetin-3-glucoside, Myricetin, Quercetin, Kaempferol, Isorhamnetin	*E. coli, S. Typhi, S. Dysenteriae, S. pneumoniae, S. aureus*	[34]
Crataegus Oxyacantha L (Rosaceae)	*Fruits*	*Ethanol Ethanol-water Methanol Methanol-water Water*	*Ultrasonic*	Naringenin Pinocembrin Kaempferol-3-glucoside	*S. aureus, E. coli, P. aeruginosa, C. albicans*	[35]
P.auriculata, F. vaillantii. F. vulgaris	*Leaves Flowers*	*Water Methanol Methanol/ water*	Maceration	Fumaric acid Carvacrol	*B. cereus, S. marcescens, E. Col, S. aureus, E. aerogenes, K. Pneumonia, P. Aeruginosa, P. Vulgaris, B. megaterium*	[30]
Ziziphus jujube	*Leaves*	*Water*	*Hot tea infusions,*	Gallocatechin, Tryptophan, Epigallocatechin, Epigallocatechin gallate, Tyrosine, kaempferol-3-O-rutinoside, Kaempferol-3-O-robinobioside	*Microganism*	[36]
C. deodara	*Pine needles*	*Water*	Maceration	Shikimic acid	*E. coli, P. vulgaris, S. aureus, B. subtilis, B. cereus*	[37]
Piper nigrum	*Black pepper*	*Chloroform Ethyl acetate Methanol Ethanol*	*Water-bath reflux (Infusions)*	Piperic acid Piperine	*E. coli, K. pneumonia, S. enterica, S. epidermidis, S. aureus, S. xylosus, B. subtilis, E. faecalis*	[38]
Momordica charantia Vine	*Vine*		*SFE*	*Plumericin*	*E. coli, P. aeruginosa, S. aureus, S. epidermidis, E. faecalis, S. typhimurium, S. mutans, B. subtilis*	[39]

Plants	Plant parts	Solvents	Methods	Compounds	Action or clinical use Microorganism	Ref.
Combretum imberbe (Combretaceae)	*Leaves*	*Dichloromethane Acetone Methanol*	Maceration	*1,2,3-dihydroxy-12-oleanen-29-oic, 1-hydroxy-12-olean-30-oic acid, 3,30-dihydroxyl-12-oleanen-22-one*	*S. aureus, E. coli*	[40]
Diospyros bipindensis	*Stem bark*	*Dichloromethane*	Maceration	*Plumbagin, canaliculatin, ismailin, betulinic acid and 4-hydroxy-5-methyl-coumarin*	*S. pneumoniae, S. aureus, S. pyogenes, E. coli, K. pneumoniae*	[41]
Anacardium occidentale	*Leaves*	*Methanol*	Maceration	*Quercetin 3-O-rutinoside Quercetin 3-O-rhamnoside*	*E coli, P. aeruginosa, S aureus, S. epidermidis, E. faecalis, S. typhimurium, S. mutans, B. subtilis*	[42]

Streptococcus pneumoniae (S. pneumoniae), Staphylococcus aureus (S. aureus), Streptococcus pyogenes (S. pyogenes), Escherichia coli (E. coli), Klebsiella pneumoniae (K. pneumoniae), Bacillus subtilis (B. subtilis), Bacillus cereus (B. cereus), Bacillus megaterium (B. megaterium), Candida albicans (C. albicans), Enterococcus faecalis (E. faecalis), Enterobacter aerogenes (E. aerogenes), Enterobacter aerogene (E. aerogenes), Listeria monocytogenes (L. monocytogenes), Pseudomonas aeruginosa (P. aeruginosa), Proteus vulgaris (P. vulgaris), Proteus vulgaris (P. vulgaris), Streptococcus mutans (S. mutans), Salmonella typhimurium (S. aureus), Staphylococcus epidermidis (S. epidermidis), Serratia marcescens (S. marcescens), Salmonella enteritidis (S. enteritidis), Yersinia enterocolitica (Y. enterocolitica); References (Ref); Microwave assisted extraction (MEA); Supercritical fluid extraction (SFE).

Accelerated Solvent Extraction (ASE) is an automated extraction technique that rapidly performs solvent extractions using high temperatures and pressure. The automation of ASE allows for high sample throughput and, if needed, fast method development. ASE is accepted by several government and regulatory agencies and produces solvent extracts with similar or better recoveries than traditional extraction techniques.

A review from Brusotti et al. [10] recently highlights the traditional solid-liquid extraction methodologies such as maceration, infusion, decoction and boiling under reflux, a wide range of modern techniques have been introduced in the past decades. These include microwave-assisted extraction (MAE), ultrasound assisted extraction (UAE), supercritical fluid extraction (SFE), and pressurized liquid extraction (PLE). Recently, these methods have been widely used in the extraction of various bioactive compounds from different plant materials [11-13].

In a recent work by Naima et al. [14], reported a comparison between the effectiveness of traditional methods (maceration and infusion) and new extraction method (microwave) for extraction of polyphenols and tannins from leaves and stem bark. Highest polyphenols yields, condensed and hydrolysable tannins were obtained by microwave followed by maceration at room temperature and infusion. In previous studies reported by Mustapa et al. [15], the comparison of MAE, SFE and soxhlet extraction of bioactive compounds from leaves and stems of C. nutans showed that microwave assisted extraction (MAE) technique was more efficient than soxhlet and supercritical fluid extraction (SFE) method for extraction of C. nutans in a shorter time with high content of antimicrobial compounds. On the other hand, Chang et al. [16] reported a comparison between MAE, UAE and conventional methodologies which highlights the advantages of MAE and UAE concerning extraction time (shorter) and extraction yield of bioactive components (higher). However, further the selection of correct extraction method is important for obtaining the extracts with the required pharmacological activity. Annegowda et al. [17] reported also a comparison between ultrasonic extraction, soxhlet and maceration and confirmed ultrasonic extraction to be an ideal, simple and rapid method to obtain antioxidant- as well as antibacterial-enriched B. purpurea leaves extract.

Choosing of Extraction Solvents

The selection of solvent system largely depends on the specific nature of the bioactive compound being targeted (Table 1). Different solvents, depending on their polarity, extract varying quantities of components in crude plant material that may be beneficial or harmful to biological systems. Up to 6 years ago most of the information about antimicrobial compounds extraction derived from the isolation and NMR identification studies. Extraction of different antimicrobial agents depends on extraction method and solvent (non-polar to polar solvents). Polar organic solvents such as aqueous solutions of methanol, ethanol or acetone, as well as ethyl acetate have been used to extract antimicrobial agent from plants (leaves, roots and barks) and they have been also the best solvents for extraction several components including phenolic acid, flavonoids, alkaloids, saponins, glycosides and hydrolysable and condensed tannins known for their antimicrobial properties [18-23]. Non-polar organic solvents (i.e., n-hexane, petroleum ether), chloroform and dichloromethane have low extraction strength for antimicrobial agent such as hydrolysable tannins, polyphenol compounds and are usually used in sample treatment to remove lipids and chlorophyll and/or to prevent enzymatic reactions [24]. Traditionally, the dried herbs are either boiled in water to make a tea used to treat systemic bacterial and fungal infections or applied directly on the skin or nails in a plaster form to treat local infections [25]. Karar et al. [23] reported quinic acid and quinic acid derivative isolated from methanol extract of quince showed a strong inhibition activity against Escherichia coli. Bhattacharjee et al. [26] also reported that ethanol was the best solvent for extracting of antibacterial compound from the plants leaves used in Indian traditional folk medicine followed by n-hexane and aqueous extracts using maceration method.

Isolation and Identification of Antimicrobial Compounds

Recently, many scientists were preoccupied by finding of appropriate methods to identify and isolate antimicrobial compounds from plants. Most studies on the bioactivity and active components of plants focused on the leaves, roots, stem bark and fruits. Antimicrobial activities of these different parts of plants are mainly due to the presence of phenolic acids, flavonoids, ellagitannins, gallotannins, and alkaloids, aglycones and saponin in their extracts. However, the combination of various types of bioactive compounds with different polarities in extracts of plants, their separation still remains a big challenge for the process of purification and identification of bioactive compounds. Their purification, identification and structural elucidation methods have always been of great interest to several scientists working in the antimicrobial field. There are several methods for the purification and identification of the bioactive compounds. These methods include chromatographic techniques: thin layer chromatography (TLC), high performance liquid chromatography (HPLC) or coupled with mass spectrometry and gas chromatography (GC). Hyphenated techniques for natural product isolation, structure elucidation, pharmaceutical analysis, impurity profiling, degradation studies and trace analysis: gas chromatography-mass (GC-MS), liquid chromatography-mass (LC-MS), liquid chromatography nuclear magnetic resonance (LC-NMR), liquid chromatography nuclear magnetic resonance mass (LC-NMR-MS) and bioassay-guided isolation. These methods have been the analytical techniques that have dominated the separation and characterization of phenolic compounds such as phenolic acids, phenylpropanoids, flavonoids and quinines.

HIGH PERFORMANCE LIQUID CHROMATOGRAPHY (HPLC)

High performance liquid chromatography (HPLC) is a versatile, robust, and widely used technique for the isolation of natural products [43]. Natural products are frequently isolated following the evaluation of a relatively crude extract in a biological assay in order to fully characterize the active entity. HPLC is ideally suited to the rapid processing of such multicomponent samples on both an analytical and preparative scale. Chemical separations can be accomplished using HPLC by utilizing the fact that certain compounds have different migration rates given a particular column and mobile phase. The extent or degree of separation is mostly determined by the choice of stationary phase and mobile phase. Generally the identification and separation of phytochemicals can be accomplished using isocratic system (using single unchanging mobile phase system). Gradient elution in which the proportion of organic solvent to water is altered with time may be desirable if more than one sample component is being studied and differ from each other significantly in retention under the conditions employed.

Purification of the compound of interest using HPLC is the process of separating or extracting the target compound from other (possibly structurally related) compounds or contaminants. Each compound should have a characteristic peak under certain chromatographic conditions. Purification on chromatography depends of the chosen conditions, such as the proper mobile phase, flow rate, suitable detectors and columns to get an optimum separation.

Characterization of the bioactive phyto-compounds in crude natural product extracts requires some form of feedback from the bioassay, which was initially used to detect the biological activity. This is necessary regardless of the separation technique and analytical method used. A common strategy has been to collect fractions from the high-performance liquid chromatography (HPLC) separation in deep-dish micro-titer plates or tubes and then resubmit the individual fractions to the original assay. This approach requires desiccation of fractions to remove the HPLC solvents, which are usually incompatible with the bioassay, resuspending the fractions in a compatible solvent (water, DMSO, or Tween), and then individual assaying of each fraction. This process is not cost effective, being both time and labor intensive.

Liquid chromatography coupled with mass spectrometry (LC/MS) is a powerful technique for the characterization of bioactive compounds of complex botanical extracts [44-46]. It provides abundant information for structural elucidation of the compounds when tandem mass spectrometry (MSn) is applied. Therefore, the combination of HPLC and MS facilitates rapid and accurate identification of chemical compounds in medicinal herbs, especially when a pure standard is unavailable [47].

HPLC – Waters Auto-Purification System

New technical are developed to isolate these compounds from complex natural products, such as HPLC-auto-purification system for the isolation and purities determination of fractions. Using the four-channel autopurification system and SPE greatly reduced the preparative time; this method was demonstrated to be reliable and efficient, especially in guaranteeing the stability of the unstable compounds and can be adopted as a new technical platform to isolate unstable compounds from complex natural products. One of the main challenges of the purification of byproducts and impurities is the low concentration of the target compounds with respect to the main peak. HPLC-autopurification system such as waters purification systems, the leader in preparative chromatography, offers scalable systems to meet your isolation and purification requirements (Figure 2). With a combination of integrated high-performance instrumentation, robust chemistry, easy-to-use software, and world renowned support services. Preparative-scale chromatography plays a critical role in applications where compounds must be synthesized, identified, isolated, purified, characterized, screened and tested. The autopurification HPLC/MS System provides you with the flexibility of using high-throughput parallel runs for selective mass-directed fraction collection from hundreds of samples. Further, the purification systems reliably isolate and purify microgram to multigram quantities of the compounds of interest. The software can trigger collection using a variety of detector signals including UV/Visible, evaporative light scattering, MS and analog.

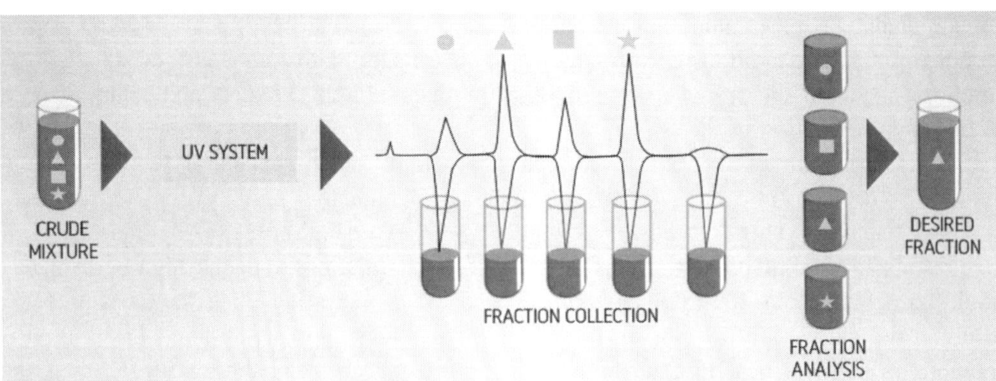

Figure 2. Isolation and purification of fractions with autopurification system.

As a first step, the samples are subject to experiments with high resolution LC-TOF-MS (Liquid Chromatography Time-of-Flight Mass Spectrometry) to gather information about molecular mass and structure. If these tests do not deliver satisfactory results, a further characterization with NMR is the next choice. Each compound should have a characteristic peak under certain chromatographic conditions. Depending on what needs to be separated and how closely related the samples are, the chromatographer may choose the conditions, such as the proper mobile phase, flow rate, suitable detectors and columns to get an optimum separation. The NMR method requires typically about one mg of substance with high purity. With the APS autopurification system's analytical option, the users develop an analytical method providing optimum resolution for all peaks to be purified within one run.

Thin-Layer Chromatography (TLC) and Bio-Autographic Methods

TLC is a simple, quick, and inexpensive procedure that gives the researcher a quick answer as to how many components are in a mixture. TLC is also used to support the identity of a compound in a mixture when the Rf of a compound is compared with the Rf of a known compound. Additional tests involve the spraying of phytochemical screening reagents, which cause color changes according to the phytochemicals existing in a plants extract; or by viewing the plate under the UV light. This has also been used for confirmation of purity and identity of isolated compounds [48]. Several methods are developed for the identification, separation and purification of bioactive compounds. TLC bioautography is an ideal, convenient and simple method for the rapid detection and separation of antioxidants as well as antimicrobials from complicated plant extract [33, 49]. Compared to other methods, the TLC bioautography method can quickly detect and separate the active components in a complicated plant extract, and has additional advantages such as convenience, being simple to run, and requiring no specialized equipment.

Bio-autography is a useful technique to determine bioactive compounds with antimicrobial activity from plant extracts. Paper chromatography followed by bioautography was used for the first time in 1946 by Goodal and Levi to estimate the purity of penicillin [50]. Thin layer chromatography - bioautography was introduced by R. Fisher and H. Lautner in 1961 [51]. There are three different approaches for bioautography to localize antimicrobial activity on a TLC chromatogram - agar diffusion or contact bioautography, direct bioautography and immersion or agar overlay bioautography [52, 53]. Those assays supply a quick screen for new antimicrobial compounds through bioassay-guided isolation [25]. TLC bioautographic methods combine chromatographic separation and in situ activity determination facilitating the localization and target-directed isolation of active constituents in a mixture. Traditionally, bioautographic technique has used the growth inhibition of microorganisms to detect antimicrobial components of extracts chromatographed on a TLC layer. This methodology has been considered as the most efficacious assay for the detection of anti-microbial compounds [1, 25]. Bio-autography localizes antimicrobial activity on a chromatogram using three approaches: (1) direct bio-autography, where the micro-organism grows directly on the thin-layer chromatographic (TLC) plate, (2) contact bio-autography, where the antimicrobial compounds are transferred from the TLC plate to an inoculated agar plate through direct contact and (3) agar overlay bio-autography, where a seeded agar medium is applied directly onto the TLC plate [54]. The inhibition zones produced on TLC plates by one of the above bioautographic technique will be use to visualize the position of the bioactive compound with antimicrobial activity in the TLC fingerprint with reference to Rf values [1].

Mass Spectrometric Application

GC-MS is suited to every organic chemical discipline where GC is applied, including the chemical, pharmaceutical, environmental, and forensics industries, as well as basic research, it is a useful technique to the structure elucidation, molecular weight determination, detection of drugs and their metabolites, or drugs of abuse in human body fluids, organic compounds and trace elements in human body fluids. Mostly, the target compounds are quantified by the external method, using respective reference compounds for establishing calibration curves.

Often, the target antibacterial components of the essential oils from plants material are identified by the external method with GC–MS method. Essential oils are volatile, natural, complex compounds that are produced by plants as secondary metabolites for protection against bacteria, viruses, fungi and pests [54]. Essential oils of aromatic plant species are used in industries for the production of soaps, perfumes and toiletries. Many of them are also used in traditional medicine for various purposes. Investigations on the evaluation of the biological activities of essential oils of some medicinal plant species have revealed that some of them exhibited interesting activities such as insecticidal, antibacterial and antifungal. Antimicrobial compound isolated and identified as xanthorrhizol [5-(1, 5-dimethyl-4-hexenyl)-2 methylphenol] from Cinnamomum Iners leaves by various spectroscopic techniques exhibited antimicrobial activity for both Gram negative and Gram positive pathogens [1, 2].

Nuclear Magnetic Resonance (NMR)

NMR spectroscopy measures the resonances of magnetic nuclei such as 1H, 13C and 15N that interact with an external magnetic field. NMR spectroscopy can be used to identify the constitution and configuration of natural products, and to characterize their various physical and chemical properties. It offers non-invasive structural analysis of metabolites in crude extracts, cell suspensions, intact tissues or whole organisms allowing in vivo analysis [56]. The NMR method provides simultaneous access to both qualitative and quantitative information since the signal intensity is directly proportional to the molar concentration.

The quality of an NMR analysis is not dependent on the number of signals detected but on the number of metabolites identified. To facilitate the identification of metabolites databases of NMR spectra of common plant metabolites are needed. The development of databases has contributed to the development of NMR, into a fast, convenient and effective metabolomic tool [57].

To enhance sensitivity selectivity and spectral resolution two-dimensional NMR is applied. Correlated spectroscopy (COSY), heteronuclear multiple quantum coherence (HMQC) spectroscopy, heteronuclear single quantum coherence (HSQC) spectroscopy and heteronuclear multiple bond coherence (HMBC) spectroscopy improve metabolite identification by providing information on the relationship between the signals from two different nuclei's [58].

NMR spectroscopy has been an important analytical tool for investigating natural compounds for many years. It is an excellent alternative to X-ray diffraction for compounds that are difficult to crystallize. Like X-ray diffraction, NMR analysis gives good quality information (e.g., composition, conformation) about the structures of simple natural compounds obtained from plants. The rapid development of NMR techniques has offered many new opportunities for more effective research on natural products.

BIOASSAY GUIDED ISOLATION

Bioassay-guided isolation integrates the processes of separation of compounds in a mixture, using various analytical methods, with results obtained from biological testing. The process begins with testing an extract to confirm its activity, followed by crude separation of the compounds in the matrix and testing the crude fractions. Further fractionation is carried out on the fractions that are found to be active at a certain concentration threshold, whereas the inactive fractions are set aside or discarded. The process of fractionation and biological testing is repeated until pure compound(s) are obtained. Reducing the number of steps may also lead to reduced loss of the interesting components. It is applied often on several bacteria and fungi whish are inherently resistant to many antimicrobial agents such as *Staphylococcus aureus, Pseudomonas aeruginosa, Escherichia coli, Salmonella typhimurium* and Aeromonas hydrophila and fungi such as *Candida albicans* and *Aspergillus niger.* Structural identification of the pure compound then follows (Figure 2). This methodology precludes overlooking novel compounds that are often missed in studies that only identify those compounds with which the investigator is familiar. Several natural compounds have been isolated by using this methodology [59, 60].

Bioassays are quantified by comparing the response of the indicator organism to an unknown concentration of a known antimicrobial agent in the sample to the responses produced by known concentrations of the same antimicrobial agent in a standard response line. A plot of logarithm of the concentration against linear inhibition zone diameter should form a straight line without obvious deviations. Deviations in linearity of a response line are most likely to occur at concentration range extremes unless technical errors were made during preparation of the solutions, during application of the solutions to the assay plates, or measurement of zone diameters. The concentration range chosen must encompass the range of concentrations expected in the sample or dilution of the sample. When determining the drug concentrations for the standard response line, it is generally appropriate to avoid concentrations that produce small (<13 mm) or large (>27 mm) zones of inhibition. When calculating the drug concentration in a specimen, do not extrapolate the concentration using inhibition zones with a diameter that exceeds the boundaries of the standard response line. The response may not be linear.

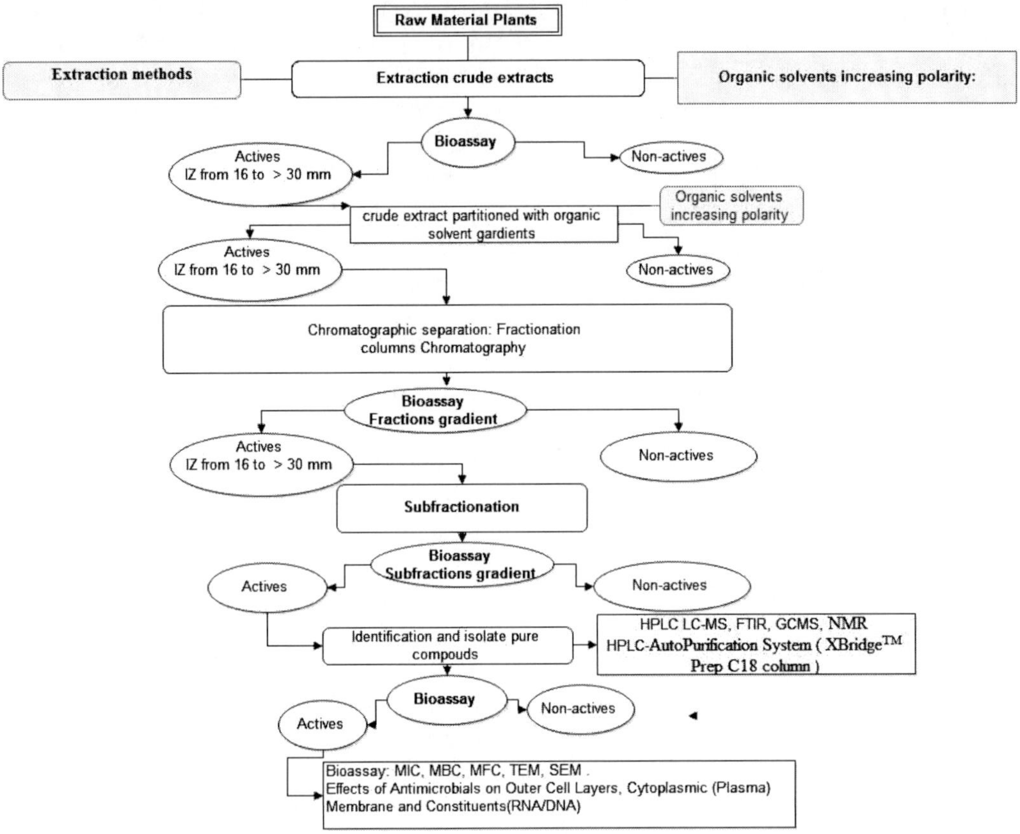

Figure 3. Preparation process of extracts and bioassay-guided isolation of antimicrobial compounds from plants.

Bioassays should only be run by well-trained technical staff. Strict control of analytical technique is necessary to minimize the likelihood of errors. Sequential carryover errors may occur when serial dilutions are used to produce standard response line concentrations. Standard response line preparation schema should minimize use of serial dilutions. Control samples of known antimicrobial content stored under the same conditions as the specimens should be tested with each batch of assays to evaluate assay quality and accuracy (Figure 3). Antibacterial bioassays typically have sensitivity in the range of 0.1 to 5 μg/mL [61].

Bioassay-guided fractionation is a procedure of whereby extract is chromatographically fractionated and refractionated until a pure biologically active compound is isolated. Each fraction produced during the fractionation process is evaluated in a bioassay system and only active fractions are fractionated. Bioassay-guided fractionation method is commonly employed in drug discovery research due to its effectiveness to directly linked the analyzed extracts and targeted compounds using fractionation procedure that followed with certain biological activity. Recently, the scientist have focalized their research in isolation process of antimicrobial activity, several antimicrobial agents have been isolated through a bioassay-guided procedure from plants such as pentacyclic triterpenoid has been isolated from the leaves of Combretum imberbe (Combretaceae) [40]. Plumbagin and ismailin were also found to be responsible for the main biological activities observed from dichloromethane stem bark extract of the *Diospyros bipindensis* [41] whereas D'Almeida et al. [62] used bioassay-guided fractionation procedure to isolate compounds: Cinnamic acid, caffeic acid methyl ester, ferulic acid methyl ester, ferulic acid ethyl ester, 3-(4-hydroxy-phenyl)-propionic acid methyl ester, 3-(3,4-dihydroxy-phenyl)-propionic acid methylester, isoferulic acid prenyl ester, ferulic acid prenyl ester, methyl ferulic acid prenyl ester, ferulic acid 2-phenethyl alcohol, dimethylcaffeic acid 2-phenethyl ester, vanillin and 4-hydroxy-3-methoxy benzoic acid methyl ester that capable of inhibit the growth of several microorganism strains from *Parastrephia lucida Cabrera.*

CONCLUSION

The question of choosing an optimal identification and isolation process for a specific bioactive compound has been the big challenge for several scientists. Although several studies have shown that the identification and isolation process of antimicrobial compounds from different plant parts, the data are still quite controversial, and little is known about the likely components associated. This chapter has been only modest reference to the older residue data about different process used to identify and isolate the antimicrobial compounds from plants [27-42].

REFERENCES

[1] Mustaffa F, Indurkar J, Ismail S, Shah M, Mansor SM. 2011. An antimicrobial compound isolated from Cinnamomum iners leaves with activity against methicillin-resistant Staphylococcus aureus. *Molecules*. 16, 3037-47.

[2] Gurib-Fakim A. 2006. Medicinal plants: traditions of yesterday and drugs of tomorrow. *Molecular aspects of medicine*. 27, 1-93.

[3] Farombi EO. 2003. African indigenous plants with chemotherapeutic potentials and biotechnological approach to the production of bioactive prophylactic agents. *African J. Biotech.* 2 662-71.

[4] VanWyk B, M. Wink. 2004. Medicinal Plants of the World. Briza Publications.

[5] Rios JL, Recio MC. 2005. Medicinal plants and antimicrobial activity. *Journal of ethnopharmacology*. 100, 80-4.

[6] Gibbons S. 2005. Plants as a source of bacterial resistance modulators and antiinfectives agents. *Phytochemistry reviews*. 4, 63-78.

[7] Adetutu A, Morgan WA, Corcoran O. 2011. Antibacterial, antioxidant and fibroblast growth stimulation activity of crude extracts of Bridelia ferruginea leaf, a wound-healing plant of Nigeria. *Journal of ethnopharmacology*. 133, 116-9.

[8] Fabricant DS, Farnsworth NR. 2001. The Value of Plants Used in Traditional Medicine for Drug Discovery. *Environmental Health Perspectives* 109, 69-75.

[9] Muniz-Marquez DB, Martinez-Avila GC, Wong-Paz JE, Belmares-Cerda R, Rodriguez-Herrera R, Aguilar CN. 2013. Ultrasound-assisted extraction of phenolic

compounds from Laurus nobilis L. and their antioxidant activity. *Ultrasonics sonochemistry*. 20, 1149-54.

[10] Brusotti G, Cesari I, Dentamaro A, Caccialanza G, Massolini G. 2014. Isolation and characterization of bioactive compounds from plant resources: the role of analysis in the ethnopharmacological approach. *J. Pharm. Biomed. Anal.* 87, 218-28.

[11] Dai J, Mumper RJ. 2010. Plant phenolics: extraction, analysis and their antioxidant and anticancer properties. *Molecules*. 15, 7313-52.

[12] Jadhav D, B. N R, Gogate PR, Rathod VK. 2009. Extraction of vanillin from vanilla pods: A comparison study of conventional soxhlet and ultrasound assisted extraction. *Journal of Food Engineering*. 93, 421-6.

[13] Hromadkova Z, Kost'alova Z, Ebringerova A. 2008. Comparison of conventional and ultrasound-assisted extraction of phenolics-rich heteroxylans from wheat bran. *Ultrasonics sonochemistry*. 15, 1062-8.

[14] Naima R, Oumam M, Hannache H, Sesbou A, Charrier B, Pizzi A, Charrier – El Bouhtoury F. 2015. Comparison of the impact of different extraction methods on polyphenols yields and tannins extracted from Moroccan Acacia mollissima barks. *Industrial Crops and Products*. 70, 245-52.

[15] Mustapa AN, Martin Á, Mato RB, Cocero MJ. 2015. Extraction of phytocompounds from the medicinal plant Clinacanthus nutans Lindau by microwave-assisted extraction and supercritical carbon dioxide extraction. *Industrial Crops and Products*. 74, 83-94.

[16] Chang CH, Yusoff R, Ngoh GC, Kung FW. 2011. Microwave assisted extractions of active ingredients from plants. *J. Chromatogr. A*. 1218, 6213-622.

[17] Annegowda HV, Mordi MN, Ramanathan S, Hamdan MR, Mansor SM. 2011. Effect of Extraction Techniques on Phenolic Content, Antioxidant and Antimicrobial Activity of Bauhinia purpurea: HPTLC Determination of Antioxidants. *Food Anal. Method*. 5, 226-33.

[18] Hatano T, Eerdunbayaer, Nozaki A, Takahashi E, Okamoto K, Ito H, Bao L-M. 2012. Hydrolysable Tannins Isolated from Syzygium aromaticum: Structure of a New C-Glucosidic Ellagitannin and Spectral Features of Tannins with a Tergalloyl Group. *Heterocycles*. 85, 365.

[19] Orabi MA, Taniguchi S, Terabayashi S, Hatano T. 2011. Hydrolyzable tannins of tamaricaceous plants. IV: Micropropagation and ellagitannin production in shoot cultures of Tamarix tetrandra. *Phytochemistry*. 72, 1978-89.

[20] Wang Y, Yang J, Wang A, Ma J, Tian J, Ji T, Su Y. 2013. Hydrolyzable tannins from Balanophora polyandra. *Acta Pharmaceutica Sinica B*. 3, 46-50.

[21] Si C-L, Xu J, Lu Y-Y, Su Z, Su Y-F, Bae Y-S. 2011. Hydrolysable tannins from Juglans sigillata stem barks. *Biochemical Systematics and Ecology*. 39, 225-7.

[22] Parveen I, Threadgill MD, Hauck B, Donnison I, Winters A. 2011. Isolation, identification and quantitation of hydroxycinnamic acid conjugates, potential platform chemicals, in the leaves and stems of Miscanthus x giganteus using LC-ESI-MSn. *Phytochemistry*. 72, 2376-84.

[23] Karar MGE, Pletzer D, Jaiswal R, Weingart H, Kuhnert N. 2014. Identification, characterization, isolation and activity against Escherichia coli of quince (Cydonia oblonga) fruit polyphenols. *Food Research International*. 65, 121-9.

[24] Mokgotho MP, Gololo SS, Masoko P, Mdee LK, Mbazima V, Shai LJ, Bagla VP, Eloff JN, Mampuru L. 2013. Isolation and Chemical Structural Characterisation of a

Compound with Antioxidant Activity from the Roots of Senna italica. *Evidence-based complementary and alternative medicine: eCAM*. 2013, 1-6.

[25] Yang X, Summerhurst DK, Koval SF, Ficker C, Smith ML, Bernards MA. 2001. Isolation of an antimicrobial compound from Impatiens balsamina L. using bioassay-guided fractionation. *Phytotherapy research: PTR*. 15, 676-80.

[26] Bhattacharjee I, Chatterjee SK, Ghosh A, Chandra G. 2011. Antibacterial activities of some plant extracts used in Indian traditional folk medicine. *Asian Pacific Journal of Tropical Biomedicine*. 1, S165-S9.

[27] Perianayagam JB, Sharma SK, Pillai KK, Pandurangan A, Kesavan D. 2012. Evaluation of antimicrobial activity of ethanol extract and compounds isolated from Trichodesma indicum (Linn.) R. Br. root. *Journal of ethnopharmacology*. 142, 283-6.

[28] Arima H, Danno G. 2002. Isolation of antimicrobial compounds from guava (Psidium guajava L.) and their structural elucidation. *Bioscience, biotechnology, and biochemistry*. 66, 1727-30.

[29] Ahmed D, Saeed R, Shakeel N, Fatima K, Arshad A. 2015. Antimicrobial activities of methanolic extract of Carissa opaca roots and its fractions and compounds isolated from the most active ethyl acetate fraction. *Asian Pacific Journal of Tropical Biomedicine*. 5, 541-5.

[30] Jaberian H, Piri K, Nazari J. 2013. Phytochemical composition and in vitro antimicrobial and antioxidant activities of some medicinal plants. *Food Chem*. 136, 237-44.

[31] Govardhan Singh RS, Negi PS, Radha C. 2013. Phenolic composition, antioxidant and antimicrobial activities of free and bound phenolic extracts of Moringa oleifera seed flour. *Journal of Functional Foods*. 5, 1883-91.

[32] Deng Y, Yang G, Yue J, Qian B, Liu Z, Wang D, Zhong Y, Zhao Y. 2014. Influences of ripening stages and extracting solvents on the polyphenolic compounds, antimicrobial and antioxidant activities of blueberry leaf extracts. *Food Control*. 38, 184-91.

[33] Soberón JR, Sgariglia MA, Sampietro DA, Quiroga EN, Sierra MG, Vattuone MA. 2010. Purification and identification of antibacterial phenolics from Tripodanthus acutifolius leaves. *Journal of applied microbiology*. 108(5), 1757–68.

[34] Yogendra Kumar MS, Tirpude RJ, Maheshwari DT, Bansal A, Misra K. 2013. Antioxidant and antimicrobial properties of phenolic rich fraction of Seabuckthorn (Hippophae rhamnoides L.) leaves in vitro. *Food Chem*. 141, 3443-50.

[35] Kostić DA, Velicković JM, Mitić SS, Mitić MN, Randelović SS. 2012. Phenolic Content, and Antioxidant and Antimicrobial Activities of Crataegus Oxyacantha L (Rosaceae) Fruit Extract from Southeast Serbia. *Tropical Journal of Pharmaceutical Research*. 11.

[36] Zhang R, Chen J, Shi Q, Li Z, Peng Z, Zheng L, Wang X. 2014. Phytochemical analysis of Chinese commercial Ziziphus jujube leaf tea using high performance liquid chromatography–electrospray ionization-time of flight mass spectrometry. *Food Research International*. 56, 47-54.

[37] Zeng WC, He Q, Sun Q, Zhong K, Gao H. 2012. Antibacterial activity of water-soluble extract from pine needles of Cedrus deodara. *International journal of food microbiology*. 153, 78-84.

[38] Zarai Z, Boujelbene E, Ben Salem N, Gargouri Y, Sayari A. 2013. Antioxidant and antimicrobial activities of various solvent extracts, piperine and piperic acid from Piper nigrum. *LWT - Food Science and Technology*. 50, 634-41.

[39] Saengsai J, Kongtunjanphuk S, Yoswatthana N, Kummalue T, Jiratchariyakul W. 2015. Antibacterial and Antiproliferative Activities of Plumericin, an Iridoid Isolated from Momordica charantia Vine. *Evidence-based complementary and alternative medicine: eCAM*. 2015, 823178.

[40] Angeh JE, Huang X, Sattler I, Swan GE, Dahse H, Hartl A, Eloff JN. 2007. Antimicrobial and anti-inflammatory activity of four known and one new triterpenoid from Combretum imberbe (Combretaceae). *Journal of ethnopharmacology*. 110, 56-60.

[41] Cesari I, Hoerle M, Simoes-Pires C, Grisoli P, Queiroz EF, Dacarro C, Marcourt L, Moundipa PF, Carrupt PA, Cuendet M, Caccialanza G, Wolfender JL, Brusotti G. 2013. Anti-inflammatory, antimicrobial and antioxidant activities of Diospyros bipindensis (Gurke) extracts and its main constituents. *Journal of ethnopharmacology*. 146, 264-70.

[42] Ajileye OO, Obuotor EM, Akinkunmi EO, Aderogba MA. 2015. Isolation and characterization of antioxidant and antimicrobial compounds from Anacardium occidentale L. (Anacardiaceae) leaf extract. *Journal of King Saud University - Science*. 27, 244-52.

[43] Cannell RJP. 1998. *Natural Products Isolation*. New Jersey: Human Press Inc. pp. 165-208.

[44] Dorta E, González M, Lobo MG, Sánchez-Moreno C, de Ancos B. 2014. Screening of phenolic compounds in by-product extracts from mangoes (Mangifera indica L.) by HPLC-ESI-QTOF-MS and multivariate analysis for use as a food ingredient. *Food Research International*. 57, 51-60.

[45] Ma Y, Kosińska-Cagnazzo A, Kerr WL, Amarowicz R, Swanson RB, Pegg RB. 2014. Separation and characterization of phenolic compounds from dry-blanched peanut skins by liquid chromatography–electrospray ionization mass spectrometry. *Journal of Chromatography A*.

[46] Jaiswal R, Karakose H, Ruhmann S, Goldner K, Neumuller M, Treutter D, Kuhnert N. 2013. Identification of phenolic compounds in plum fruits (Prunus salicina L. and Prunus domestica L.) by high-performance liquid chromatography/tandem mass spectrometry and characterization of varieties by quantitative phenolic fingerprints. *J. Agric. Food Chem*. 61, 12020-31.

[47] Ye M, Han J, Chen H, Zheng J, Guo D. 2007. Analysis of phenolic compounds in rhubarbs using liquid chromatography coupled with electrospray ionization mass spectrometry. *J. Am. Soc. Mass. Spectrom*. 18, 82-91.

[48] Sasidharan S, Chen Y, Latha LY. 2011. Extraction, isolation and characterization of bioactive compounds from plants' extracts. *Afr. J. Tradit. Complement. Altern. Med*. 8 (1), 1-10.

[49] Gu L, Wu T, Wang Z. 2009. TLC bioautography-guided isolation of antioxidants from fruit of Perilla frutescens var. acuta. *LWT - Food Science and Technology*. 42, 131-6.

[50] Goodall RR, Levi AA. 1946. A microchromatographic method for the detection and approximate determination of the different penicillins in a mixture, *Nature*. 158, 675-6.

[51] Betina V. 1973. Bioautography in paper and thin-layer chromatography and its scope in the antibiotic field. *J. Chromatography*. 78, 41-51.

[52] Patil NN, Waghmode MS, Gaikwad PS, Gajbhai MH, Bankar AV. 2013. Bioautography guided Screening of Antimicrobial Compounds Produced by Microbispora V2. *International Research Journal of Biological Sciences*. 2(2), 65-8.

[53] Cos P, Vlietinck AJ, Berghe DV, Maes L. 2006. Anti-infective potential of natural products: How to develop a stronger in vitro 'proof-of-concept.' *J. Ethnopharmacol.* 106, 290-302.

[54] Ejele AE, Akalezi CI, Iwu IC, Ukiwe LN, Enenebaku CK, Ngwu SU. 2014. Bioassay-Guided Isolation, Purification and Charaterization of Antimicrobial Compound From Acidic Metabolite of Piper umbellatum Seed Extract. *International Journal of Chemistry*. 6.

[55] Rota MC, Herrera A, Martínez RM, Sotomayor JA, Jordán MJ. 2008. Antimicrobial activity and chemical composition of Thymus vulgaris, Thymus zygis and Thymus hyemalis essential oils. *Food Control*. 19, 681-7.

[56] Fan T-M, Lane AN. 2008. Structure-based profiling of metabolites and isotopomers by NMR. *Progr. Nucl. Magn. Reson. Spectr.* 52, 69-117.

[57] Verpoorte R, Choi Y, Mustafa N, Kim H. 2008. Metabolomics: back to basics. *Phytochemistry reviews: proceedings of the Phytochemical Society of Europe*. 7, 525-37.

[58] Leiss KA, Choi YH, Verpoorte R, Klinkhamer PG. 2011. An overview of NMR-based metabolomics to identify secondary plant compounds involved in host plant resistance. *Phytochemistry reviews: proceedings of the Phytochemical Society of Europe*. 10, 205-16.

[59] Chakraborty D, Chakraborti S. 2010. Bioassay- Guided Isolation and identification of antibacterial and anti fungal components from methanolic extract of Green tea leaves (Camellia sinensis). *Research Journal of Phytochemistry*. 4(2), 78-86.

[60] Maryam J, Ihsan uH, Bushra M, Mazhar Q. 2012. Isolation of antibacterial compounds from Quercus dilatata L. through bioassay guided fractionation. *Annals of clinical microbiology and antimicrobials*. 11, 11.

[61] Donald HP, Estrella M-M. Bioassay Methods for Antimicrobial and Antifungal Agents. *Antimicrobial Susceptibility Testing Protocols: CRC Press* 2007:313-40.

[62] D'Almeida RE, Alberto MR, Quispe C, Schmeda-Hirschmann G, Isla MI. 2012. Antimicrobial phenylpropanoids from the Argentinean highland plant Parastrephia lucida (Meyen) Cabrera. *Journal of ethnopharmacology*. 142, 407-14.

In: High Value Processing Technologies
Editors: A. Berenjian, H. Jafarizadeh-Malmiri et al.
ISBN: 978-1-63484-453-6

Chapter 9

PRODUCTION OF POLYUNSATURATED FATTY ACIDS BY OLEAGINOUS MICROORGANISMS AND PLANTS

***Xin Tang*[1] *and Yuanda Song*[1,2,*]**
[1]State Key Laboratory of Food Science and Technology,
School of Food Science and Technology, Jiangnan University, Wuxi, Jiangsu, China
[2]Colin Ratledge Center for Microbial Lipids,
School of Agriculture Engineering and Food Science,
Shandong University of Technology, Zibo, Shandong, China

ABSTRACT

Polyunsaturated fatty acids (PUFAs) have critical biological functions but mammals lack the ability to synthesize some PUFAs, therefore they must be supplied in the diet. In the recent years, traditional breeding and fermentation technologies for PUFAs production have been developed and some PUFAs (e.g., γ-linolenic acid, arachidonic acid, docosahexaenoic acid) have been commercially produced by oleaginous microorganisms. Futhermore, the transgenic plants and yeasts are the appropriate sources to produce PUFAs. Overexpression of various desaturases in plant significantly enhanced its PUFAs production, and the integration of some PUFAs biosynthetic pathways in transgenic yeast made it able to accumulate large amount of PUFAs. The fermentation technology and trasgenic technology are very important in PUFAs production by oleaginous microoraganisms or plants, and the commercial production of various PUFAs will become ture with the fast development of modern molecular biotechnology.

Keywords: polyunsaturated fatty acids, oleagious microorganisms, plants, transgenic

[*] Corresponding author: Yuanda Song, Colin Ratledge Center for Microbial Lipids, School of Agriculture Engineering and Food Science. Shandong University of Technology. Email: ysong@sdut.edu.cn.

INTRODUCTION

Fatty acid biosynthesis in most organisms culminates in the formation of either C16 or C18 saturated fatty acids. These fatty acids are then modified through a sequence of desaturases and elongases so that an extended range of unsaturated and polyunsaturated fatty acids (PUFAs) are produced [1]. PUFAs, with their unique structural and functional characteristics, are distinguished by two main functions. The first function relates to their roles in regulating the architectures, dynamics, phase transition and permeability of membranes, and modulating the behavior of membrane-bound proteins. The other function is that PUFAs control the expression of certain genes and thus affect some processes including fatty acid biosynthesis and cholesterol transport in the body [2, 3]. Much more interests are focused on the second role of PUFAs as precursors of a wide variety of metabolites (e.g., prostaglandins, leukotrienes and hydroxy-fatty acids) regulating critical biological functions. The various roles played by PUFAs make it apparent that PUFAs deficiencies lead to abnormalities in the skin, nervous system, immune and inflammatory systems, cardiovascular system, endocrine system, kidneys, respiratory and reproductive systems [4]. Since mammals lack the ability to synthesize some PUFAs, they must be supplied in the diet [5].

TRADITIONAL BREEDING AND FERMENTATION TECHNOLOGY FOR PUFAS PRODUCTION

The first PUFA that was produced commercially was γ-linolenic acid (GLA, 18:3; n-6). GLA is a naturally occurring PUFA with recognized beneficial effects for prevention and treatment of inflammatory disorders, diabetes, cardiovascular disorders, cancers, atopic dermatitis and some other diseases [6, 7]. The main natural source of GLA has been seed oils from plants, such as evening primrose (8-10% GLA in total fatty acids), but the low production rate made it expensive. As an alternative, an industrial fermentation system, using fungi (e.g., *Mortierella isabellina*, *Mortierella ramanniana* and *Mucor circinelloides*), was developed [8]. In these organisms, GLA is biosynthesized from a precursor, linoleic acid (LA, C18:2; $\triangle$9,12), through the introduction of a double bond at the $\triangle$6 position of the carbon backbone. Development of a biotechnological route to produce an equivalent microbial lipid rich in GLA has been described in detail. This process, which was in commercial production in the United Kingdom from 1985 to 1990, was the world's first microbial lipids to be offered for sale, using the filamentous fungus, *Mcr. circinelloides* (18% GLA in total fatty acids) [7]. Indeed, it was superior to evening primrose oil in all respects (e.g., higher content of GLA, higher stability to oxidation, absence of high levels of competing fatty acid, lower content of herbicide and pesticide residues). Although the fungal oil contained about twice the concentration of GLA as did evening primrose oil, it experienced, perhaps not surprisingly, some marketing problems (e.g., the source from fungi was not accept well by the market). Moreover, the price of evening primrose was deliberately decreased so as to become more competitive and, simultaneously, borage oil was developed as an improved plant source of GLA [9].

After GLA, arachidonic acid (ARA, 20:4n-6) was the next PUFA to be developed and the development processes for its production has been conducting until today. ARA is incorporated into infant feed formula where it is considered a desirable component to be added along with docosahexaenoic acid (DHA). Traditional sources of ARA are egg yolks and animal livers. The former source yields various phospholipids rich in ARA that are then used as such while the latter source, although containing ARA as a triacylglycerol, is not acceptable vegetarians [9]. Microbial sources of ARA being actively developed have mainly used *Mort. alpina* as the best producing organism. Currently, large scale processes exist in both Japan and Europe for its production. In particular, studies have indicated positive effects on infant cognitive development after ARA supplementation and an International Expert Group have advised the inclusion of PUFAs such as ARA and DHA in infant formula [10]. Additionally, ARA deficiency is characterized by hair loss, fatty liver degeneration, anaemia and reduced fertility in adults although such deficiencies are not common in adults given the relatively high consumption of ARA, especially in Western diets. At a cellular level, ARA is a component of cell membranes and is also a precursor to eicosanoids, a family of molecules which have broad roles in regulating inflammatory responses, blood clotting and cell signaling [11]. ARA is an essential fatty acid for many mammals and is synthesized by a number of micro-organisms including many microalgae, thraustochytrids, fungi and some bacteria [12-14]. The most common synthesis pathway appears to commence with a △6-desaturation of LA to produce GLA which is subsequently acted on by a △6-elongase to produce dihomo-γ-linolenic acid (DGLA, 20:3; △8,11,14) and finally a △5-desaturase to yield ARA. An alternative pathway exists, however, which commences with a △9-elongation of LA to produce eicosadienoic acid (EDA, 20:2; △11,14) and is followed by a △8-desaturation to form DGLA which can then be △5-desaturated to produce ARA as by the △6-desaturase pathway.

Other PUFAs that are being produced by, or are capable of production by, microorganisms are eicosapentaenoic acid (EPA, 20:5n-3) and docosahexaenoic acid (DHA, 22:6n-3). These two PUFAs occur together in the oils of many fishes, and recommendations for their use as dietary supplements for the prevention of cardiac problems in older people have long been advocated. Stocks of many fish species are now dwindling rapidly and the increasing presence of pollutants in the marine environment cause concern for the future and safe supplies of these fatty acids. EPA performs many vital functions in biological membranes and serves as a precursor of a variety of lipid regulators in cellular metabolism. Therefore it plays an important role in the treatment of various human diseases, such as rheumatoid arthritis, heart disease, cancers, schizophrenia, and bipolar disorder [15-17]. These findings have led to considerable interest in developing commercial processes to produce EPA [18, 19]. Marine fish oil is currently the richest source for EPA and is currently used for its commercial production; however, EPA recovery and purification from fish oils is expensive [18]. Microorganisms, including microalgae, lower fungi and marine bacteria, are the primary producers of EPA; fish usually get EPA from bioaccumulation in the food chain [20, 21]. Much effort is being devoted to develop a commercially feasible technology to produce EPA directly from microorganisms. The biosynthesis of EPA occurs through a series of reactions that can be divided into two distinct steps. First is the de novo synthesis of oleic acid (OA, 18:1n-9) from acetate. This is followed by conversion of OA to LA and α-linolenic acid (ALA, 18:3n-3), and a number of subsequent stepwise desaturation and elongation steps to form an n-3 PUFAs family that includes EPA.

DHA is essential for the growth and functional development of the brain in infants [22]. DHA is also required for maintenance of normal brain function in adults [23]. The inclusion of plentiful DHA in the diet improves learning ability, whereas deficiencies of DHA are associated with deficits in learning [24]. DHA is taken up by the brain in preference to other fatty acids and the turnover of DHA in the brain is very fast, much more rapid than that is generally realized. The visual acuity of healthy, full-term, formula-fed infants is increased when their formula includes DHA [23]. DHA deficiency is associated with foetal alcohol syndrome, attention deficit hyperactivity disorder, cystic fibrosis, phenylketonuria, unipolar depression, aggressive hostility, and adrenoleukodystrophy and decreases in DHA in the brain are associated with cognitive decline during aging and with onset of sporadic Alzheimer disease [25]. The leading cause of death in western nations is cardiovascular disease, while epidemiological studies have shown a strong correlation between fish (contains abundant DHA) consumption and reduction in sudden death from myocardial infarction [22]. However, the DHA to be produced in an oil without the presence of EPA, which is thought to compete against DHA for incorporation into key retinal and neural brain lipids, is required. As it is very difficult to remove EPA from the mixture of DHA and EPA in fish oils, this has promoted to develop microorganisms as sources of DHA as species are known that produce this PUFA without the presence of EPA. There are two principal organisms currently used commercially for production of an oil rich in DHA: *Crypthecodinium cohnii* and the thraustochytrid group of marine microorganisms [26]. The former is a marine dinoflagellate that has been known for some time as a DHA producer. Although this particular PUFA is rare in oils from any source and it is uncertain by which route it may be synthesized, it does not seem to be deletrious to the efficacy of DHA in its incorporation into key membrane lipids of the human body. Numerous species may be capable to synthesize DHA, few of them have been identified as being suitable producers from an economic point. In particular, if considering heterotrophic production strains that can be grown in conventional stirred fermenters rather than phototrophic strains that require either open pond systems or sophisticated photobioreactor technology, only a few candidates remain. One of these is the dinoflagellate, *C. cohnii*, another one is *Schizochytrium sp.*, that belongs to the group of Thraustochytriales [27, 28]. Both these organisms have been described extensively as potential DHA producers and are being used today for commercial DHA production [26, 29, 30].

TRANSGENIC PLANTS FOR PUFAS

GLA can be produced from its precursor LA by the action of a △6 desaturase, but this reaction is not common in nature. Notably, among higher plants such as borage, evening primrose, and blackcurrant have been reported to use this route to accumulate GLA [9, 31]. However, it is not appropriate to mass market supply as these species are not part of the modern diet, moreover, they are difficult to cultivate and within relatively low oil yields. Therefore there is an opportunity for transforming an oleaginous source for GLA production. The first published experiments reported that an expression of a △6 desaturase isolated from the cyanobacterium *Synechocystis sp.* in *Nicotiana tabacum* (tobacco) plants, resulting in a low level of GLA accumulation in transgenic leaves, but not in seeds [32]. Whilst much higher levels of GLA accumulation were obtained by changing the △6 desaturase to one from *B. officinalis* [33]. The expression of the *B. officinalis* △6 desaturase under the control of a constitutive promoter in *Linium ustatissimum* (linseed) also resulted in the accumulation of GLA [34]. Taken together, these early research demonstrated the viability of using transgenic approaches to modify seed oil PUFAs content. Considerable progress has been made and further studies have shown very high levels (up to 70%) of GLA as a result of seed-specific expression. It was reported that expression of a *B. officinalis* △6 desaturase in *Brassica juncea*, can produce 9% GLA in total fatty acids [34]. Liu et al. expressed a fungal △6 desaturase along with a △12 desaturase from *Mort. alpina*, producing up to 43% GLA in seeds of *Brassica napus* [35]. Similar levels of GLA were produced in *B. juncea* seeds when expressing a △6 desaturase from the fungus *Pythium irregular* [36]. More recently, the seed oil of a high LA cultivated species of safflower (*Carthamus tinctorius*) was modified by transformation with △6 desaturase from *Saprolegnia diclina*, resulting in levels exceeding 70% GLA content [37].

As the genetic biotechnology of producing PUFAs in plant had developed, longer PUFAs such as EPA and DHA was also produced in transgenic plant. The appropriate host plant has become more critical as the characteristics of their own genes related to PUFA biosynthesis can affect the PUFA production. *Brassica carinata* seemed to be a better host species for EPA production than *B. juncea*, producing an 20% EPA in transgenic seeds, while, in both species, the high levels of GLA in transgenic seeds indicated the inefficiency of the Δ6 elongation. Only when EPA production in seeds reach an appreciable level, production of DHA will be realistically attempted. The first report of DHA synthesis in oilseeds was carried out by Dupont [38]. They demonstrated the expression of △6 desaturase, △5 desaturase and △6 elongase from the fungus *Mort. alpina* in soybean. Individual genes were under the control of different seed-specific promoters and both a n-3 △17-desaturase from *S. diclina* and a n-3 △15-desaturase from *Arabidopsis* were included so that maximize the accumulation of n-3 PUFAs by turning the n-6 PUFA metabolites into n-3 metabolites. In this case, they achieved EPA levels of 9% in somatic soybean embryos. Replacing the *Mort. alpina* △6 desaturase with an *S. diclina* △6 desaturase appeared to slightly increase EPA levels in embryos (EPA levels up to 13%) and plants derived from some of these embryos produced seeds with EPA levels reaching up to almost 20%. However, these embryos also produced up to 4% docosapentaenoic acid (DPA; 22:5 △7,10,13,16,19), the elongation product of EPA, demonstrating that the *Mort. alpina* elongase had some activity on C20 fatty acids when expressed in plants. Therefore, in order to produce DHA in somatic embryos, a △4 desaturase from *Schizochytrium aggregatum* and a specific △5 elongase from *Pavlova sp.* were also expressed. Unexpectedly, no more than 4% of DHA in total fatty acids was achieved and these low yields were presumably a result of a poor acyl exchange between the △5 elongase and the △4 desaturase.

The entire DHA synthesis in *B. juncea* was reconstituted using modern molecular biotechnology [39]. The first construct only contained the minimal set of genes required to produce ARA and EPA, and they are △6 desaturase from the fungus *P. irregular*, △6 elongase from *P. patens*, and △5 desaturase from *Thraustochytrium*. The transgenic seeds contained high levels of GLA (27%), but relatively low levels of C20 PUFA, thus indicating the poor performance of the reactions required for △6 elongation. Nevertheless, these seeds also contained 7% ARA and 0.8% EPA resulting from the high conversion efficiency of the △5 desaturase. As *B. juncea* oil contains abundant LA which is a n-6 substrate in the PUFA synthesis, the seeds of transgenic plants accumulated much higher amounts of the n-6 fatty acid ARA than the n-3 fatty acid EPA. In the second construct, a *Calendula officinalis* △12 desaturase was added to ensure an increase in LA. This experiment resulted in notably higher levels of ARA (18%) as well as EPA (1.3%). The next step was the addition of a second △6-elongase from *Thraustochytrium spp.* to this construct, and it only produced a low increase in elongation efficiency. The final addition of a *Phytophtora infestans* n-3 △17 desaturase produced oil that contained higher levels of EPA (8%) with a concomitant reduction in ARA. In order to synthesize DHA in *B. juncea* seeds, three extra genes to the transformation vector: a △6/△5-elongase from the fish species *Oncorhynchus mykiss* within a △4 desaturase and a putative lysophosphatidic acid acyltransferase from *Thraustochytrium sp.* were expressed [39]. It is conceivable that the acyltransferase from a DHA-rich organism might improve the exchange between different pools of intermediate substrates. However, these experiments suggested the acyltransferase activity did not always contribute to the increased synthesis of long PUFAs. The transgenic seeds showed an accumulation of 8.1% EPA while the highest value was 15%. However, there was a poor conversion of EPA into DPA which was rate limiting for the △4 desaturase producing DHA.

The biosynthesis of EPA and DHA in transgenic plants has been clearly demonstrated by a number of groups, however, the levels of long PUFAs are not as high as those seen in fish oil [40]. Nowadays, the aim is to generate a vegetable oil like as fish oils optimizing the accumulation of long PUFAs and avoiding the accumulation of intermediates. In order to do this, the regulation, organization, and assembly of the elongase complex must be improved. Furthermore, it is also important that the variation in long PUFA levels could be partly due to the different promoters [41].

TRANSGENIC YEASTS FOR PUFAS

Yarrowia lipolytica, as a oleaginous yeast, has the considerable potential for producing PUFAs using modern molecular biotechnology since its nice growth characteristics (e.g., efficiently grow on numerous carbon sources) together with the availability of the genomic information and good genetic tools for gene manipulation [42]. Conjugated linoleic acid (CLA) is geometric and positional dienoic isomers of linoleic acid (18:2, n-9) and has been extensively studied because of its health benefits including cancer prevention, anti-atherogenic and anti-obesity effects, and immune system modulation [43]. Zhang et al. expressed linoleic acid isomerase gene from *Propionibacterium acnes* (PAI) in *Y. lipolytica* to produce CLA [44]. Codon usage optimization of the PAI and multi-copy integration significantly improved the expression level of PAI, and the percentage of trans-10, cis-12 CLA was six times higher in yeast carrying the codon-optimized gene than that carrying the native gene. In combination with multi-copy integration, the production yield was raised to approx. 30-fold. The amount of trans-10, cis-12 CLA reached 5.9% of total fatty acid yield in transformed *Y. lipolytica*. Later, they go on co-expressing △12 desaturase gene from *Mort.*

alpina together with the codon-optimized linoleic acid isomerase (oPAI) gene in *Y. lipolytica* [45]. The genetically-modified *Y. lipolytica* produced trans-10, cis-12-CLA at a level of up to 10% of total fatty acids and 0.4% of cell dry weight when using glucose as the sole carbon source. However, when the recombinant yeast was grown with soybean oil, trans-10, cis-12-CLA accumulated at a level of up to 44% of total fatty acids, which represented 30% of cell dry weight. Furthermore, trans-10, cis-12-CLA was also detected in the growth medium up to 0.9 g/l. Another application of transgenic *Y. lipolytica* is using the genes metabolic biotechnology to producing EPA. Xue et al. reported the engineering of the oleaginous yeast *Y. lipolytica* for commercial production of EPA [46]. Metabolic engineering of the oleaginous yeast *Y. lipolytica* resulted in a strain that produced EPA at 15% of dry cell weight. The engineered yeast lipid comprises EPA at 56.6% and saturated fatty acids at less than 5% by weight, which are the highest and the lowest percentages, respectively, among known EPA sources. Inactivation of the peroxisome biogenesis gene *PEX10* was crucial in obtaining high EPA yields and may increase the yields of other commercially desirable lipid-related products. This technology platform enables the production of lipids with tailored fatty acid compositions and provides a sustainable source of EPA. In addition, an engineered strain of the oleaginous yeast *Y. lipolytica* capable of producing greater than 5.6% DHA in the total oil fraction is also constructed by Dupont [47]. This strain comprises various chimeric genes expressing heterologous desaturases, elongases and acyltransferases and optionally comprises various native desaturase and acyltransferase knockouts to enable synthesis and high accumulation of DHA.

FUTURE WORKS AND PROSPECT

The use of microbial lipids as new food applications to produce PUFAs will depend on the application, commodity pricing of alternatives, and delivery form of the microbial lipids. The use of the whole microbial lipids biomass, without the additional extraction costs and losses, may offer an economic opportunity for functional nutrition in the animal field. A number of patents have been filed on the use of microbial lipids biomass in animal feeds to improve the DHA content of the animal or a part of that is to be consumed. Examples include feeding of *Schizochytrium* biomass to chickens in order to produce high-DHA eggs [48], or feeding cows with a protected fish-oil product to produce DHA-enriched milk [49]. Such concepts are limited by the extent of bioconversion of the feed DHA into the DHA content of the salable food product, without incurring substantial losses in general metabolism, conversion into other unused body parts, or losses in the feces. The commercial development of microbial lipids has gone through a series of ups and downs since the first recognition of their potential over 100 years ago. Initially microbial lipids were academic curiosities. Then, there were several attempts at commercialization, but each failed due to lower cost alternatives on the market. Nowadays, with the fast development of modern molecular biotechnology, the commercial production of DHA and ARA using microbial has been realized.

REFERENCES

[1] Ratledge, C. (2004). Fatty acid biosynthesis in microorganisms being used for single cell oil production. *Biochimie* 86, 807-815.

[2] Horrobin, D. F. and Manku, M. S. (1983). Essential fatty acids in clinical medicine. *Nutrition and Health* 2, 127-134.

[3] Sessler, A. M. and Ntambi, J. M. (1998). Polyunsaturated fatty acid regulation of gene expression. *The Journal of Nutrition* 128, 923-926.

[4] Holman, R. T. (2009). Polyunsaturated fatty acids. *Methods of biochemical analysis* 4, 99.

[5] Certik, M. and Shimizu, S. (1999). Biosynthesis and regulation of microbial polyunsaturated fatty acid production. *Journal of bioscience and bioengineering* 87, 1-14.

[6] Horrobin, D. (1992). Nutritional and medical importance of gamma-linolenic acid. *Progress in lipid research* 31, 163-194.

[7] Rateldge, C. (2005). *Handbook of functional lipids*, (CRC press).

[8] Suzuki, O. (1988). Production of gamma-linolenic acid by fungi and its industrialization.

[9] Ratledge, C. and Wynn, J. P. (2002). The biochemistry and molecular biology of lipid accumulation in oleaginous microorganisms. *Advances in applied microbiology* 51, 1-51.

[10] Koletzko, B., Baker, S., Cleghorn, G., Neto, U. F., Gopalan, S., Hernell, O., Hock, Q. S., Jirapinyo, P., Lonnerdal, B. and Pencharz, P. (2005). Global standard for the composition of infant formula: recommendations of an ESPGHAN coordinated international expert group. *Journal of pediatric gastroenterology and nutrition* 41, 584-599.

[11] Kinsella, J. E., Lokesh, B., Broughton, S. and Whelan, J. (1990). Dietary polyunsaturated fatty acids and eicosanoids: potential effects on the modulation of inflammatory and immune cells: an overview. *Nutrition* (Burbank) 6, 24-44.

[12] Volkman, J., Jeffrey, S., Nichols, P., Rogers, G. and Garland, C. (1989). Fatty acid and lipid composition of 10 species of microalgae used in mariculture. *Journal of Experimental Marine Biology and Ecology* 128, 219-240.

[13] Nichols, D. S., Brown, J. L., Nichols, P. D. and McMeekin, T. A. (1997). Production of eicosapentaenoic and arachidonic acids by an Antarctic bacterium: response to growth temperature. *FEMS microbiology letters* 152, 349-354.

[14] Lewis, T. E., Nichols, P. D. and McMeekin, T. A. (1999). The biotechnological potential of thraustochytrids. *Marine Biotechnology* 1, 580-587.

[15] Nettleton, J. A. (1995). Omega-3 fatty acids and health, (Springer).

[16] Gill, I. and Valivety, R. (1997). Polyunsaturated fatty acids, part 1: occurrence, biological activities and applications. *Trends in biotechnology* 15, 401-409.

[17] Peet, M. (2004). Nutrition and schizophrenia: beyond omega-3 fatty acids. *Prostaglandins, leukotrienes and essential fatty acids* 70, 417-422.

[18] Belarbi, E. H., Molina, E. and Chisti, Y. (2000). A process for high yield and scaleable recovery of high purity eicosapentaenoic acid esters from microalgae and fish oil. *Enzyme and Microbial Technology* 26, 516-529.

[19] Grima, E. M., Belarbi, E.-H., Fernández, F. A., Medina, A. R. and Chisti, Y. (2003). Recovery of microalgal biomass and metabolites: process options and economics. *Biotechnology advances* 20, 491-515.

[20] Yazawa, K. (1996). Production of eicosapentaenoic acid from marine bacteria. *Lipids* 31, S297-S300.

[21] Barclay, W., Meager, K. and Abril, J. (1994). Heterotrophic production of long chain omega-3 fatty acids utilizing algae and algae-like microorganisms. *Journal of applied phycology* 6, 123-129.

[22] Horrocks, L. A. and Yeo, Y. K. (1999). Health benefits of docosahexaenoic acid (DHA). *Pharmacological Research* 40, 211-225.

[23] Innis, S. M. (2007). Dietary (n-3) fatty acids and brain development. *The Journal of Nutrition* 137, 855-859.

[24] Innis, S. M. (2008). Dietary omega 3 fatty acids and the developing brain. *Brain research* 1237, 35-43.

[25] Quinn, J. F., Raman, R., Thomas, R. G., Yurko-Mauro, K., Nelson, E. B., Van Dyck, C., Galvin, J.E., Emond, J., Jack, C. R. and Weiner, M. (2010). Docosahexaenoic acid supplementation and cognitive decline in Alzheimer disease: a randomized trial. *Jama* 304, 1903-1911.

[26] de Swaaf, M. E., de Rijk, T. C., Eggink, G. and Sijtsma, L. (1999). Optimisation of docosahexaenoic acid production in batch cultivations by Crypthecodinium cohnii. *Journal of biotechnology* 70, 185-192.

[27] Jiang, Y. and Chen, F. (2000). Effects of temperature and temperature shift on docosahexaenoic acid production by the marine microalge Crypthecodinium cohnii. *Journal of the American Oil Chemists' Society* 77, 613-617.

[28] Kamlangdee, N. and Fan, K. (2003). Polyunsaturated fatty acids production by Schizochytrium sp. isolated from mangrove. *Songklanakarin J. Sci. Tech.* 25, 643-650.

[29] Yaguchi, T., Tanaka, S., Yokochi, T., Nakahara, T. and Higashihara, T. (1997). Production of high yields of docosahexaenoic acid by Schizochytrium sp. strain SR21. *Journal of the American Oil Chemists' Society* 74, 1431-1434.

[30] Nakahara, T., Yokochi, T., Higashihara, T., Tanaka, S., Yaguchi, T. and Honda, D. (1996). Production of docosahexaenoic and docosapentaenoic acids bySchizochytrium sp. isolated from Yap Islands. *Journal of the American Oil Chemists' Society* 73, 1421-1426.

[31] Barre, D. E. (2001). Potential of evening primrose, borage, black currant, and fungal oils in human health. *Annals of nutrition and metabolism* 45, 47-57.

[32] Reddy, A. S. and Thomas, T. L. (1996). Expression of a cyanobacterial Δ6-desaturase gene results in γ-linolenic acid production in transgenic plants. *Nature biotechnology* 14, 639-642.

[33] Sayanova, O., Smith, M. A., Lapinskas, P., Stobart, A. K., Dobson, G., Christie, W. W., Shewry, P. R. and Napier, J. A. (1997). Expression of a borage desaturase cDNA containing an N-terminal cytochrome b5 domain results in the accumulation of high levels of Δ6-desaturated fatty acids in transgenic tobacco. *Proceedings of the National Academy of Sciences* 94, 4211-4216.

[34] Qiu, X., Hong, H., Datla, N., MacKenzie, S. L., Taylor, D. C. and Thomas, T. L. (2002). Expression of borage Δ6 desaturase in Saccharomyces cerevisiae and oilseed crops. *Canadian journal of botany* 80, 42-49.

[35] Liu, J., Huang, Y., DeMichele, S., Bergana, M., Bobik, E., Hastilow, C., Chuang, L., Mukerji, P. and Knutzon, D. (2001). Evaluation of the seed oils from a canola plant genetically transformed to produce high level of-linolenic acid. Linolenic Acid: Recent Advances in Biotechnology and Clinical Applications. AOCS Press, Champaign, IL, 61-71.

[36] Hong, H., Datla, N., Reed, D. W., Covello, P. S., MacKenzie, S. L. and Qiu, X. (2002). High-level production of γ-linolenic acid in Brassica juncea using a Δ6 desaturase from Pythium irregulare. *Plant physiology* 129, 354-362.

[37] Nykiforuk, C. L., Shewmaker, C., Harry, I., Yurchenko, O. P., Zhang, M., Reed, C., Oinam, G. S., Zaplachinski, S., Fidantsef, A. and Boothe, J. G. (2012). High level accumulation of gamma linolenic acid (C18: 3Δ6. 9, 12 cis) in transgenic safflower (Carthamus tinctorius) seeds. *Transgenic research* 21, 367-381.

[38] Kinney, A., Cahoon, E., Damude, H., Hitz, W., Liu, Z.-B. and Kolar, C. (2004). Production of very long chain polyunsaturated fatty acids in oilseed plants (Google Patents).

[39] Wu, G., Truksa, M., Datla, N., Vrinten, P., Bauer, J., Zank, T., Cirpus, P., Heinz, E. and Qiu, X. (2005). Stepwise engineering to produce high yields of very long-chain polyunsaturated fatty acids in plants. *Nature biotechnology* 23, 1013-1017.

[40] Ruiz-López, N., Sayanova, O., Napier, J. A. and Haslam, R. P. (2012). Metabolic engineering of the omega-3 long chain polyunsaturated fatty acid biosynthetic pathway into transgenic plants. *Journal of experimental botany* 63, 2397-2410.

[41] Chen, R., Matsui, K., Ogawa, M., Oe, M., Ochiai, M., Kawashima, H., Sakuradani, E., Shimizu, S., Ishimoto, M. and Hayashi, M. (2006). Expression of Δ6, Δ5 desaturase and GLELO elongase genes from Mortierella alpina for production of arachidonic acid in soybean [Glycine max (L.) Merrill] seeds. *Plant science* 170, 399-406.

[42] Zhang, H., Zhang, L., Chen, H., Chen, Y. Q., Chen, W., Song, Y. and Ratledge, C. (2014). Enhanced lipid accumulation in the yeast Yarrowia lipolytica by over-expression of ATP: citrate lyase from Mus musculus. *Journal of biotechnology* 192, 78-84.

[43] Crumb, D. and Vattem, D. (2011). Conjugated linoleic acid (CLA)-An overview. *International Journal of Applied Research in Natural Products* 4, 12-15.

[44] Zhang, B., Rong, C., Chen, H., Song, Y., Zhang, H. and Chen, W. (2012). De novo synthesis of trans-10, cis-12 conjugated linoleic acid in oleaginous yeast Yarrowia lipolytica. *Microb. Cell Fact.* 11, 51.

[45] Zhang, B., Chen, H., Li, M., Gu, Z., Song, Y., Ratledge, C., Chen, Y. Q., Zhang, H. and Chen, W. (2013). Genetic engineering of Yarrowia lipolytica for enhanced production of trans-10, cis-12 conjugated linoleic acid. *Microb. Cell Fact.* 12, 70.

[46] Xue, Z., Sharpe, P. L., Hong, S.-P., Yadav, N. S., Xie, D., Short, D. R., Damude, H. G., Rupert, R. A., Seip, J. E. and Wang, J. (2013). Production of omega-3 eicosapentaenoic acid by metabolic engineering of Yarrowia lipolytica. *Nature biotechnology* 31, 734-740.

[47] Damude, H. G., Gillies, P. J., Macool, D. J., Picataggio, S. K., Ragghianti, J. J., Seip, J. E., Xue, Z., Yadav, N. S., Zhang, H. and Zhu, Q. Q. (2014). Docosahexaenoic acid producing strains of Yarrowia lipolytica (Google Patents).

[48] Herber, S. and Van Elswyk, M. (1996). Dietary marine algae promotes efficient deposition of n-3 fatty acids for the production of enriched shell eggs. *Poultry Science* 75, 1501-1507.

[49] Wright, T., Holub, B., Hill, A. and McBride, B. (2003). Effect of combinations of fish meal and feather meal on milk fatty acid content and nitrogen utilization in dairy cows. *Journal of dairy science* 86, 861-869.

In: High Value Processing Technologies
Editors: A. Berenjian, H. Jafarizadeh-Malmiri et al.
ISBN: 978-1-63484-453-6

Chapter 10

FUNGAL CELLULASE BASED ON MUTANT *ASPERGILLUS* SPICES

***Nahideh Jafari*[1], *Hoda Jafarizadeh-Malmiri*[1,*], *Maryam Hamzeh-Mivehruod*[2] *and Mohammad Adibpour*[3]**
[1]Faculty of Chemical Engineering,
Sahand University of Technology, Tabriz, Iran
[2] Biotechnology Research Center and Faculty of Pharmacy,
Tabriz University of Medical Science, Tabriz, Iran
[3]Faculty of Medicine, Tabriz University of Medical Science,
Tabriz, Iran

ABSTRACT

Agricultural wastes, which are accumulated every year in large quantities, contain mainly lignocellulose compounds, which in turn, can be converted into energy and products with high commercial interest. Cellulase has been receiving attention in recent years, due to its ability to bioconversion of lignocellulosics to useful products. Among the microorganisms, many filamentous fungi are known to produce cellulase. Most of the commercial cellulases have been produced from the *Trichoderma*, *Penicillium* and *Aspergillus* genera. However, these strains produce a little amount of cellulase, by mutation and genetically modification of them, high levels of cellulase can be obtained. Many species of Aspergillus have been used commercially to produce cellulase. *Aspergillus niger* is a fermenting microorganism, which is a well-known efficient producers of cellulolytic enzymes, through both solid-state and submerged fermentation processes. This chapter provides the current knowledge on cellulase production by *Aspergillus spp* especially *Aspergillus niger*, its industrial applications and some strain improvement methods to increase the production of fungal cellulase enzyme.

Keywords: cellulase, *Aspergillus niger*, microbial enzyme, fermentation, mutation

[*] E-mail: h_jafarizadeh@sut.ac.ir, h_jafarizadeh@yahoo.com, Tel: +98-413-3459099, Fax: +98-413-3444355.

INTRODUCTION

Lignocellulosic wastes including agricultural, agro-industrial and forestry wastes are abundant, inexpensive energy sources and renewable. These wastes contain several different components such as sugarcane bagasse, sawdust, poplar trees, brewer's spent grains, waste paper, switch grass, and straws, husks, stalks, stems, leaves, shells and peels from cereals like corn, rice, barley, wheat and sorghum, among others. A large amount of lignocellulose wastes are accumulated every year which in turn, cause ecological problems, health issues and safety hazards, and prevent sustainable development in terms of resource recovery and recycling of waste materials. However, these wastes could be utilized for the production of a number of useful products, such as bioethanol, food additives, organic acids and enzymes. The major components of lignocellulose are cellulose, hemicellulose, and lignin. Cellulose forms a skeleton of different parts of plants surrounded by hemicellulose and lignin [1, 2].

Cellulose is a linear polysaccharide of glucose residues connected by β-1,4 linkages. It is the most abundant organic source and renewable biopolymer of food, fuel and chemicals. Cellulose is found in large amounts in nearly all plants, and is potentially a major food source. Unfortunately, human beings lack the enzymes necessary to cleave the linkages between the sugars in cellulose. In fact, crystallite cellulose is added to some foods to reduce the caloric value. By development of economically feasible process technologies based on enzymatic hydrolysis, these cellulosic materials can be successfully convert to useful low molecular weight products such as hexoses and pentoses [3]. Cellulolytic enzymes refer to a group of enzymes which break down cellulose into glucose [4]. Several applied studies on cellulase have revealed its high potential to use in many different industries including agriculture, food, textile, detergents, pulp and paper and biotechnology for bioethanol production [5, 6]. The main sources of cellulases are some fungi and a few bacteria and actinomycete species. However, filamentous fungi are preferred for cellulase production via fermentation. Besides *Aspergillus species*, other fungi like *Trichoderma*, *Penicillium* and *Humicola* species are the primary sources of commercial production of cellulase [5]. Microorganisms have been utilized in industrial microbiology and biotechnology to make new products with high economical value or improve the product properties. Industrial microbiology has been used for producing agents such as antibiotics, steroids and hormones (pharmaceutical and medicinal compounds), chemical feed stocks, solvents, organic acids, amino acids and enzymes [7]. Enzymes have been widely used in several industrial processes in the areas of biotechnology [6, 8]. Considering the future views of cellulase in biotechnological applications, the main objective of this review chapter is to describe a summary of cellulase production using *Aspergillus* species, especially *Aspergillus niger*, and to present an overview of cellulase fermentation techniques and its main substrate, lignocellulose. Furthermore, this review will discuss the cellulose applications and characteristics as well as the mutation processes applied to develop cellulase production yields.

ENZYME

Enzymes are the catalysts responsible for cell metabolism. As compared to the chemical catalysts, which display only limited selectivity, most enzymes are specific. Enzymes catalyze reactions by forming a complex with their substrates at their active site (Figure 1). Generally, enzymes can be extracted or recovered from living organism cells or their exudates. Plant tissues and animal organs are the main sources of enzymes. In 1960, about 70% of the enzymes were obtained from plant tissues or exudates and animal organs. Nowadays, the most industrial enzymes are produced from microbial sources [5]. Animal derived enzymes such as catalase (EC 1.11.1.6) obtained from liver, lipase (EC 3.1.1.3), chymotrypsin (EC 3.4.21.1), and trypsin (EC 3.4.21.4) from pancreas and rennin from calf abomasum, represent about 10% of the total enzyme market. Plant derived enzymes, such as papain (EC 3.4.22.2), obtained from the latex of papaya, and bromelain, extracted from pineapple stems, are still industrially relevant and they roughly represent 5% of total market [6-11]. Since the early 1960s, microbial enzymes have gradually and progressively replaced those from other sources [3]. In a conservative estimate, they might now represent almost 90% of the total market. As compared with plant and animal, microorganisms are preferred for enzyme production due to some reasons described below [12]. Enzymes obtained from animal and plant sources require specific raw materials with careful planned collection and storage. While, microbes grow on a wide range of substrates and therefore, their enzyme production potential is huge. In microbial enzyme production, inducer molecules also can be added to fermenters to enhance enzyme production. The microbes can also be treated with mutagens (chemical or physical) to produce minor changes in enzyme structure that can result in altered enzyme characteristics. Enzymes may be produced intracellulary or extracellulary. The major difference between isolating an intracellular and extracellular enzyme is related to the cell lysis. In the case of intracellular enzyme, enzyme extraction using a suitable technique which ensures maximum cell disruption with minimum enzyme damage and purification is much important processes. The majority of enzymes in industrial use are extracellular proteins from fungal or bacterial resources making it easier and cheaper to purify such enzymes for commercial use [8, 11].

MICROBIAL ENZYMES

The microbial enzymes are used in processes as biocatalysts to perform reactions in an economical and environmentally friendly way as compared to the chemical catalysts.

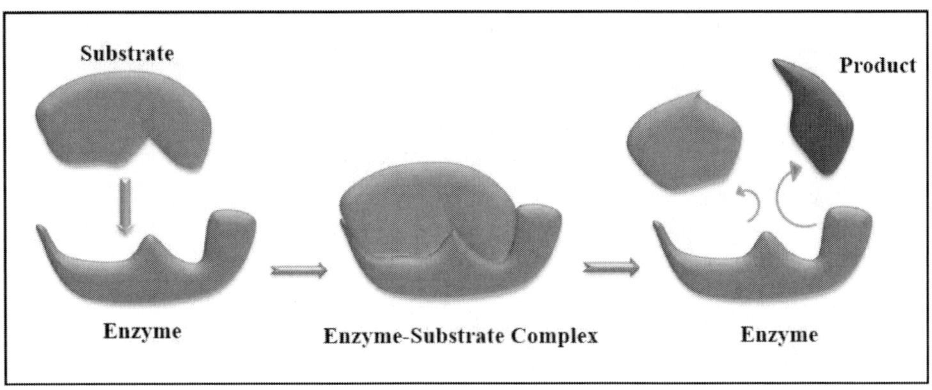

Figure 1. Mechanism of enzyme action.

Thermotolerance, tolerance to a varied range of pH, thermophilic nature, stability of enzyme activity over a range of temperature and pH, and other harsh reaction conditions are the special characteristics of enzymes which make them suitable for commercial and industrial applications [13]. Many enzymes from microbial sources are already being used in various commercial processes. Selected microorganisms including bacteria, fungi, actinomycetes and yeasts have been widely studied for the synthesis of various enzymes [6]. However, fungi are the most important microorganisms to produce commercial and industrial enzymes as compared with the other types of microorganisms [4, 14].

Application of Microbial Enzymes

Microbial enzymes have been extensively used in food, pharmaceutical, textile, paper, leather, and other industries and are being increasing dramatically [5]. The more important current uses of these enzymes are listed in Table 1.

Cellulase as an Important Industrial Enzyme

Cellulases are one the most important enzymes for industrial uses. The hydrolytic action of cellulases on cellulose, a linear polysaccharide polymer composed of many glucose units, creates free monosaccharides not only for liquid fuel production, but also for the production of valuable chemicals and bioproducts [6]. Considering the future perspectives of rising prices of crude oil due to increasing fuel demands and its limited sources, the basic and applied studies are needed to find alternative resources of bioenergy in the coming years. Among the potential alternative bioenergy resources, lignocellulosics have been known as the main source of biofuels and other value-added products. The primary components of lignocellulosic biomass, cellulose and hemicellulose, are the most abundant organic compounds on earth which make them as an excellent renewable source for energy and chemicals [52].

Cellulase is a complex enzyme system including three hydrolytic enzymes namely: endo-(1, 4)-β-D-glucanase [endoglucanase (EG), endocellulase, CMCase (EC 3.2.1.4)], exo-(1,4)-β-D-glucanase [cellobiohydrolase, exocellulase, microcrystalline cellulase, avicelase (EC 3.2.1.91)] and β-glucosidase [cellobiase (EC 3.2.1.21)] [53].

Table 1. A summarized overview of some microbial enzymes with important industrial applications

Enzyme	Producer microorganism	Industrial applications	Reference
Pectinase	*Bacillus; Erwinia, Kluyveromyces; Aspergillus; Rhizopus; Trichoderma; Pseudomonas; Penicillium; Fusarium*	Food industries (for fruit ripening, viscosity clarification and reduction of fruit juices); Textile and paper industries(fiber degumming); Oil extraction; Baby food(protein enrichment)	[15],[16], [17],[18]

Enzyme	*Producer microorganism*	Industrial applications	Reference
Amylase	*Bacillus sp.; Geobacillus; Aspergillus sp.; Rhizopus sp; Endomycessp; Mucorpusillus; Trichoderma viride*	Starch industry (for liquefaction); Paper, Food industry (Glucose and Maltose syrups, High fructose corn syrups, clarified fruit-juices); Pharmaceutical industries (Digestive aid); Brewing Industry (Starch processing); Textile industry (Warp-sizing of fibers); Baking industry (delayed staling)	[6],[19], [20],[21], [22],[23], [24]
Protease	*Bacilli; Pseudomonas; Clostridium; Rhizopus; Penicillium; Aspergillus; Mucormiehei*	Washing Powders; Detergents; Tannery; Food industry; Leather processing; Pharmaceuticals; Molecular biology; Peptide synthesis	[6],[25], [26],[20], [27],[28]
Ligninase	*Basidiomycetes strains— Steccherinumochraceum, Polyporusversicolo; Panustigrinus*	Denim washing; Bio-sensors; Bio-bleaching of kraft-pulp; Bioremediation; Pollution-control; Treatment of recalcitrant chemicals in Textile and industrial effluents	[29],[30], [31],[32], [33]
Cellulase	*Basidiomycetes strains Polyporus sp.; Pleurotus sp.; Trichoderma sp.; Aspergillus sp.*	Food industry (extraction of components from green tea, soy protein, essential oils, aromatic products and sweet potato starch.)Glucose feedstock from cellulose; Bio-refinery; Bioethanol; Paper-pulp industry; Cosmetics; Detergent and textile industries; Medical pharmaceutical industry	[34],[35], [36],[37], [38],[39], [40],[41]
Xylanase	*Thermoactinomyce sthalophilus; Bacillus sp.; Humicolainsolens; Bispora (acidophilic fungus)*	Pentose production - bioconversion of hemicellulose for fuel and Chemicals; Fruit-juice clarification; Improving rumen digestion; Paper industry- selective removals of xylans from Kraft-pulp; Brewing industry	[42],[43], [44],[45], [46]
Lipase	*Pseudomonas cepaci; Clostridium thermocellum Bacillus; Yeasts and Fungal strains Candida sp.; Aspergillus sp; Penicillium sp., Rhizopus; Mucor*	Detergents; Dairy industry oils; fats; Butter; Cream; Fat-spreads; Feed supplement; Therapeutic agent	[24],[47], [48],[49], [50],[51]

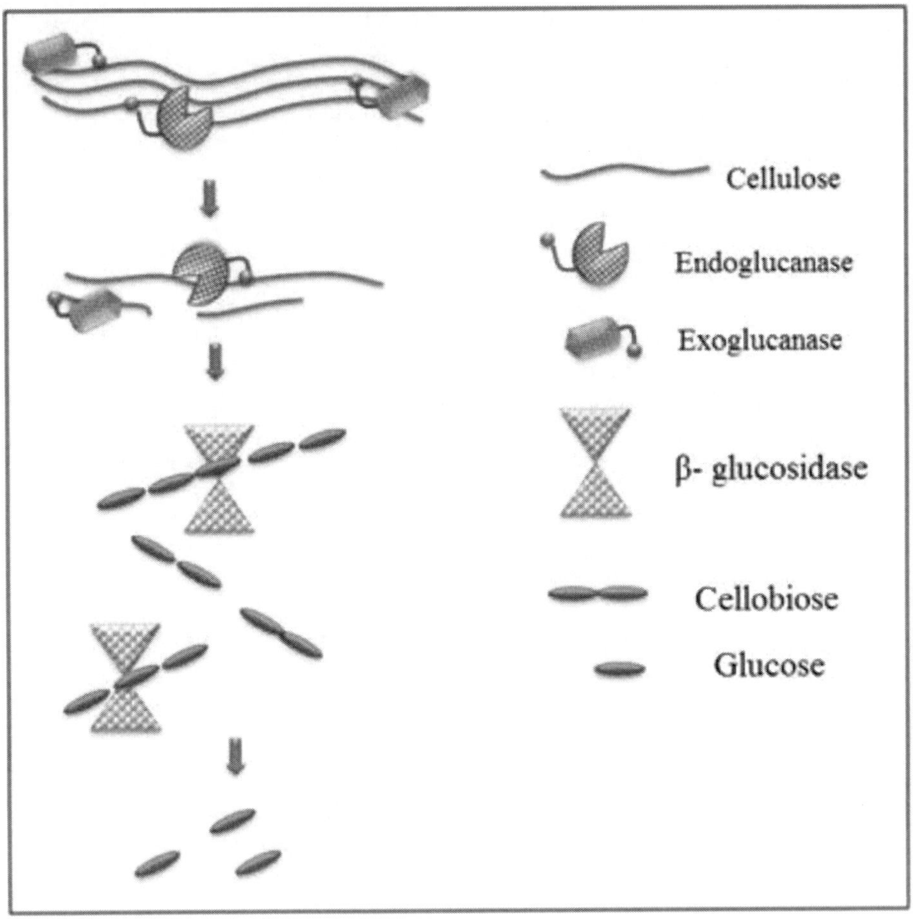

Figure 2. Mechanism of cellulase action.

In fact, these enzymes collaborate to cellulose hydrolysis by acting endoglucanases randomly along the chain, producing new attack sites for cellobiohydrolases (exoenzymes) which, releases cellobiose as their main product. Cellobiose is a strong inhibitor of endo- and exoglucanases, and the accumulation of cellobiose will slow down the hydrolysis process significantly.

β-glucosidases, which are not regarded as real cellulases, completes the process through hydrolysis of cellobiose and other short oligosaccharides to glucose [52]. The resulted glucose can be used in different applications such as production of bioethanol, single cell protein, feed stock, industrially important bioproducts [54]. Figure 2 schematically shows the hydrolytic action of cellulose.

Application of Cellulase

Through developing for several decades, the use of cellulases has considerably increased, demonstrating their massive biotechnological potential in various industries. Along with the extensive fundamental researches of cellulases, their application studies have sharply developed. Studies on cellulases began in early 1980s, first in animal feed in food applications. Subsequently, they were used in the textile, laundry as well as in the pulp and paper, agriculture, brewery industries (Table 1).

In recent years, fundamental and applied researches on cellulase enzyme have not only generated significant scientific knowledge but also have exposed their enormous potential in biotechnology. Great attention has been dedicated to the application of cellulose in bioconversion of biomass into bioethanol, as the cleanest liquid fuel and alternative to fossil fuels [53].

Microorganisms Producing Cellulase

The cellulase production has been reported from a wide variety of bacteria, actinomycetes and fungi [55]. Fungus population, such as the genera *Trichoderma*, *Chaetomium*, *Penicillium*, *Aspergillus*, *Fusarium* and *Phoma*; aerobic bacteria, such as *Acidothermus*, *Bacillus*, *Celvibrio*, *pseudonoma*, *Staphylococcus*, *Streptomyces* and *Xanthomonas*; and anaerobic bacteria, such as *Acetovibrio*, *Bacteroides*, *Butyrivibrio*, *Caldocellum*, *Clostridium*, *Erwinia*, *Eubacterium*, *Pseudonocardia*, *Ruminococcus* and *Thermoanaerobacter* have been used in cellulose production [56]. However, filamentous fungi are preferred for commercial enzyme production, due to higher enzyme production amounts of them as compared to that of the yeast, bacteria and other microorganisms [55]. Structurally fungal cellulases are also simpler as compared to bacterial cellulase systems. Fungal cellulases contain two discrete domains namely: a catalytic domain (CD) and a cellulose binding module (CBM). CBM is joined by a short poly linker region, which is rich in serine and threonine, to the catalytic domain at the N-terminal. The CBM contains of almost 35 amino acids [57]. Major fungi employed in cellulase production are listed in Table 2.

Several species of the fungal genus *Trichoderma* and *Aspergillus*, due to their high ability to secrete cellulose degrading enzymes have been widely studied and applied. These strains can be produced extracellular cellulolytic enzymes, namely endoglucanases (EC 3.2.1.4), exoglucanases (EC 3.2.1.91), β-glucosidase (EC 3.2.1.21) which acts synergistically in the conversion of cellulose to glucose. Furthermore, in order to produce high levels of cellulose, these strains could be mutagenized and genetically modified.

Trichoderma is one of the most important fungi which have been mainly used in cellulose production. In fact, most of the cellulolityc enzymes, which they degrade cellulosic components has obtained from *Trichoderma* species. Species of *Trichoderma* can produce substantial amounts of endoglucanase and exoglucanase but very low levels of β-glucosidase [87].

In most of the cellulases produced by cellulolytic fungi, such as in *Trichoderma* cellulases, the amount of β-glucosidase is lower than needed for the efficient cellulose to glucose hydrolysis and the main product of the hydrolysis is cellobiose. While, glucose as carbon source can be easily metabolized by all microorganisms as compared to the cellobiose.

β-glucosidase has several applications in food technology as well. It can degrade the most common pigments such as anthocyanins into sugars and anthocyanidins. Thus this enzyme is appropriate for decolorization of different drinks and can be used in the processing of juices like orange, and rosé [62].

Table 2. Some fungal strains used in production of cellulase

Fungal strain	Reference
Aspergillus niger	[36],[38],[39],[53],[55],[58],[59],[60], [61],[62],[63],[64],[65]
Aspergillus flavus	[66],[67],[37]
Aspergillus terreus	[68], [69],[70]
Aspergillus fumigatus	[71],[72]
Trichoderma longi	[60]
Trichoderma harziamm	[73]
Trichoderma reesei	[62],[74],[75],[76],[77],[78],[79]
Trichoderma viride	[80],[81],[82],[83]
Penicillium spp	[84],[71],[85]
Alternaria spp	[86]

Because of low β-glucosidase produced by *Trichoderma*, attention has recently been diverted to other microorganisms including the members of genus *Aspergillus* and it has widely been exploited as it possess all the three essential components of cellulase system [87].

Microbial Cellulase Production under Fermentation

Fermentation is a biological technique that can be used in large scale to convert of complex substrates into simple compounds by various microorganisms. It has been widely used for the production of enzymes such as cellulase. This technique, due to its economic and environmental advantages, has gained much of attention over the years. Two broad fermentation techniques have emerged as a result of this rapid development namely; submerged fermentation (Smf) and solid state fermentation (SSF) [88].

Submerged Fermentation (Smf)

In Smf, liquid substrates, such as molasses and broth, are utilized and suitable for microorganisms those need high moisture content (e.g., bacteria). Industrially important enzymes have traditionally been produced using Smf, due to its easier handling and greater control of environmental factors such as temperature and pH [88].

Solid State Fermentation (SSF)

SSF utilizes moist solid substrates, like bran, paddy straw, bagasse, other agricultural waste and paper pulp. The main advantage of using these substrates is that nutrient-rich waste materials can be easily recycled as cheaper substrates. SSF is best suited for fermentation techniques involving fungi and microorganisms that require less moisture content. SSF offers advantages over liquid cultivation especially for fungal cultures because the filaments of the fungi can penetrate deeply into the substrate leadind to the increasing of surface area however, in the case of bacteria, it is restricted only to the outer surface of the substrate. As compared to Smf, SSF could reduce the cost of downstream processing, utility of simple and cheap media for the fermentation, lower risk of contamination due to the ability of contaminants to grow in absence of free flowing water, high concentration of the product and simple fermentation equipment as well as low effluent generation and low requirements for aeration and agitation during enzyme production [6, 52, 55, 88].

LIGNOCELLOLUSIC SUBSTRATES

Lignocelluloses are abundant sources of carbohydrate, continually replenished by photosynthetic reduction of carbon dioxide with sunlight energy. Thus they are the most promising feedstock for the production of energy, food and chemicals. The major constituents of lignocellulose are cellulose, hemicellulose, and lignin, polymers that are closely associated with each other constituting the cellular complex of the vegetal biomass. Basically, cellulose forms a skeleton which is surrounded by hemicellulose and lignin.

Lignocellulosic wastes can be converted into products that are of commercial interest such as ethanol, glucose and single-cell protein. The inherent properties of native lignocellulosic materials make them resistant to enzymatic attack which in turn, necessitates pretreatment of the wastes before use. Since lignocellulosic materials are very complicated, their pretreatment is not simple either. The best method and conditions of pretreatment depend greatly on the type of lignocelluloses [1]. When lignocellulose is the carbon source for cellulolytic enzyme production, different pretreatment methods are used to make the substrate more available to the fungi. Steam treatment is a common method of pretreatment, but treatment with NaOH and H_2SO_4 or combinations of these methods is also used. Figure 3 shows the effect of pretreatment on lignocellulosic materials. Several lignocellulosic substrates such as wheat bran, wheat straw, rice straw, corn cob, cotton flower shell, groundnut shell wheat and sorghum straw and water hyacinth blend saw dust have been used to cellulose production [89]. Some of lignocellulosic wastes are presented in Figure 4.

CELLULASE ACTIVITY ASSAY

Several methods have been proposed for assay of cellulose activity. Sharrock (1988) presented cellulase activity assay according to two basic approaches. First approach was related to determine the activity of individual cellulases (endoglucanases, exoglucanases, and β-glucosidases) while, the second one was measuring the total saccharifying activity of a crude cellulase system by application of filter paper (filter paper assay, FPA) [90]. Zhang et al. (2006) categorized all cellulase activity assays into three main groups based on monitoring and measurement of the accumulation of products after hydrolysis [92]. Due to the complexity of cellulose-cellulase systems and differences between kinetic characteristics of initial hydrolysis reaction and the extended time, cellulase activity assays are either expressed based on the initial hydrolysis rate or using the end-point hydrolysis.

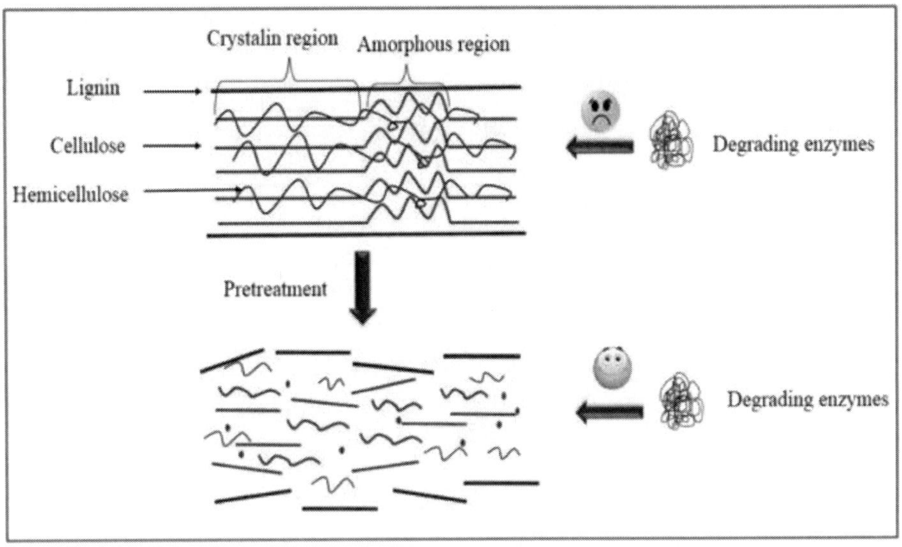

Figure 3. Effect of pretreatment on accessibility of degrading enzymes.

Figure 4. Different types of Lignocellulosic wastes.

In fact, for measuring an individual cellulase activity in a short time, the first assay method is preferred and for the total enzyme activity assay within a given time, the last one is used [91-93]. FPA is the key method for analysis of total cellulase activity (i.e., three cellulase enzymes). In 1976, the filter paper assay was developed by Mendel's et al. [94]. FPA is the most common total cellulase activity assay recommended by the International Union of Pure and Applied Chemistry (IUPAC) [93].

Carboxymethylcellulase activity (CMCase) or Endoglucanases (EG) can randomly hydrolyze internal glycosidic bonds in cellulose chains. EGs activities can be measured using a soluble cellulose derivative with a high degree of polymerization (DP) such as carboxymethylcellulose (CMC). CMCase is mainly evaluated based on the procedure described by Mendel's et al. [94]. Cellobiohydrolases (exoglucanases) or Avicellulases are classified as exo-acting based on the assumption that they all cleave β-1, 4-glycosidic bonds from chain ends releasing cellobiose and some glucose molecules. for measuring exoglucanase activity, commercial Avicel (also called microcrystalline cellulose or hydrocellulose) is utilized due to its low degree of polymerisation of cellulose and relatively inaccessible to attack by endoglucanases despite some amorphous regions [93]. β-glucosidase activity can be measured using various chromogenic and nonchromogenic substrates and are mainly evaluated based on the procedure of Kubicek [95].

CELLULASE ENZYME PRODUCED BY *ASPERGILLUS SPP*

Approximately, 200 species of *Aspergillus* have potential of enzyme production. They are isolated from soil, decomposing plants and air, and produce a great number of extracellular enzymes, which can be widely applied in biotechnology. Some of the most important species of the genus *Aspergillus* are; *A.flavus, A. niger, A. oryzae, A. nidulans, A. fumigatus, A. clavatus, A. glaucus, A. ustus* and *A. versicolor* [56].

Aspergillus niger is an important soil fungi having a great potential of producing a range of primary and secondary metabolites like citric acid gluconic acid, cellulases and other enzymes. Moreover, *Aspergillus niger* is one of the few species of the fungus kingdom classified as GRAS (Generally Recognized as Safe) by the Food and Drug Administration (FDA).

Aspergillus filamentous fungi stand out as major producers of cellulolytic enzymes. Several researches have indicated that the filamentous fungus *Aspergillus niger,* can be produced cellulolytic enzymes, organic acids and other valuable bioproducts via solid state fermentation processes [56]. Its microscopic features are shown in Figure 5. Most of fungi of genus *Aspergillus* can synthesis cellulase, therefore, this genus has the potential to dominate the enzyme industry [87]. Table 3 demonstrates the previous reports on *Aspergillus spp.* producing cellulose.

CHALLENGES IN MICROBIAL PRODUCTION OF CELLULASE

The bioconversion of cellulosic materials has been receiving much attention in recent years. It is now a subject of intensive research to the development of a large scale conversion process beneficial to mankind [101]. Such process as suggested by researchers would help alleviate shortages of food and animal feeds, solve modern waste disposal problem, and diminish man's dependence on fossil fuels by providing a convenient and renewable source of energy in the form of glucose [101-103]. But the cost of production and low yield of these enzymes are the major problems for their industrial applications [38].

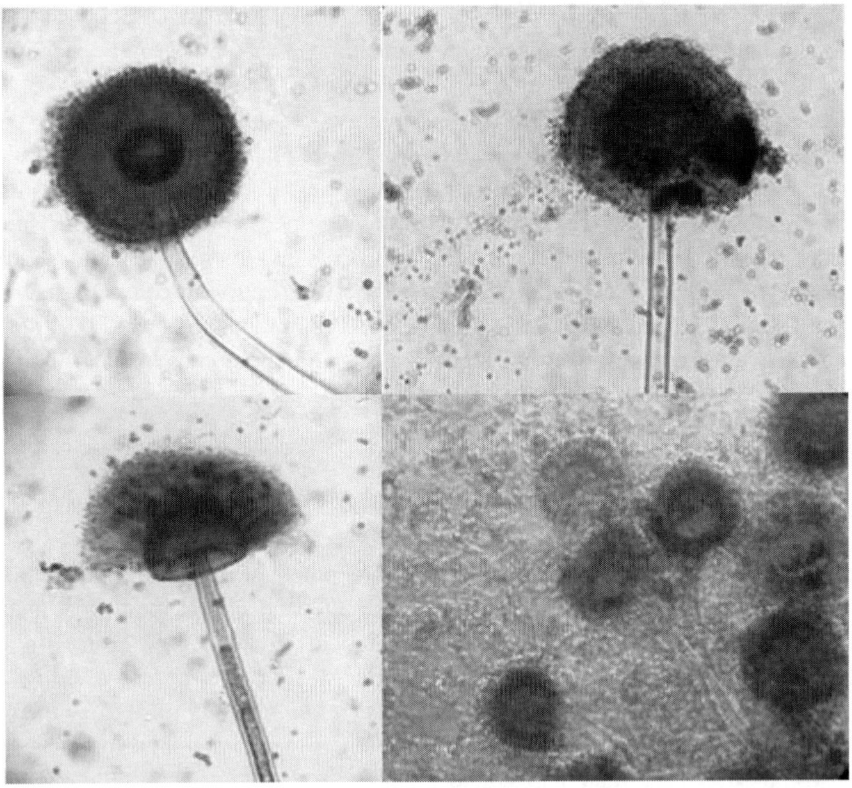

Figure 5. Microscopic features of *Aspergillus niger* stained with lactophenol cotton blue at magnifications of 40X.

Thus, low yield and high operating costs of commercially available cellulase has led to extensive research into producing high yielding and active fungal cellulase enzymes. The exponential increase in the application of enzymes in various fields in the last few decades demands extension in both qualitative improvement and quantitative enhancement through strain improvement and medium optimization for higher enzymatic yield. Improvement of microbial strains for the over-production of industrial products has been the trademark of all commercial fermentation processes.

Strain Improvement

The science and technology of manipulating and improving microbial strains, to increase their metabolic capacities, which in turn, enhances biotechnological applications, are defined as strain improvement. Strain improvement, can be achieved by two strategies, namely: classical genetic and molecular genetic methods [104].

Table 3. Cellulase produced using *Aspergillus spp*

Fungi	Substrate	Technique	Enzyme	Enzyme activity	Ref.
A. flavus	Sawdust, bagasse, corncob	Smf	Cellulase	0.0743IU/ml 0.0573IU/ml 0.0502IU/ml	[37]
A. niger, T. reesei	Waste paper	Smf	β-glucosidase	3.07 IU/mL	[62]
A. niger	Rice straw	SSF	CMCase, Fpase, β-glucosidase	5.9IU/ml 0.62 IU/ml 1.11 IU/ml	[63]
A. niger	Coir waste	SSF Smf	CMCase, Fpase	3.29 U/ml, 2.3 U/ml	[55]
Fungi	Substrate	Technique	Enzyme	Enzyme activity	Ref.
A. niger	Wheat straw Rice straw	Smf	CMCase	0.577 ± 0.13 IU 0.456 ± 0.11 IU	[96]
A. niger	Banana peel powder, coir powder	SSF	Cellulase	0.072 ±0.001 IU/ml 0.046 ±0.003 IU/ml	[65]
A. niger	Rice husk Saw dust	Smf	Cellulase	13.702±0.1 IU 9.683±0.2 IU	[97]
A. niger	CMC	SSF	CMCase	14.122 U/mg protein	[53]
A. niger	Paper, timber sawmill	SSF	Cellulase	3.9 IU	[64]
A. niger	Rice bran, Orange peel, Carboxymethylcellulose	Smf	Cellulase	16.63µmol/min 14.30µmol/min 3.61µmol/min	[98]
A. niger	Cellulose-supplemented media Sawdust	Smf	Cellulase	0.54 units mg $Protein^{-1}$ 0.30 units mg $Protein^{-1}$	[99]
A. terreus	Oil palm empty fruit bunch	SSF	Fpase, CMCase, β-glucosidase	0.76 U/ml 8.64 U/ml 6.81 U/ml	[69]
A. terreus	Bagasse	Smf	CMCase	2.43 U/ml	[68]
A. terreus	Groundnut shell	Smf	CMCase exoglucanase	1.147 IU/ml 0.175 FPU/ml	[100]

In this chapter, classical genetic methods will be discussed. Strain development by this method is basically depended on mutation, followed by random screening and selection of new improved mutants via fermentation tests.

Classical Genetic Methods

Mutation

Mutation is a very important process which leads to variations in genes by altering in the sequence of nitrogenous bases of a DNA molecule. Mutagenizes is the process of producing a mutation and mutagen is the agents which cause mutation. Mutagens are natural or synthetic agents which can alter the structure or sequence of DNA. When a mutagen is used in mutation, the resulting mutation is called induced mutation. While, if mutation is occurs spontaneously in nature without a known mutagen, it is called spontaneous mutagenesis and the resulting mutation is spontaneous mutation or basal mutation. [105]. Different mutagenic agents which are classified into the physical and chemical mutagens, can be utilized to increase induced mutation rates [106]. In this review chapter, physical mutagens are highlighted.

Physical Mutagens

Radiations

The different types of radiations having mutagenic properties are known as physical mutagens. The radiations may be a part of the electromagnetic spectrum having shorter wavelength and higher energy than visible light (UV rays, X rays, gamma rays and cosmic rays) or may be particulate radiations produced by the decay of radio isotopes [105, 107]. Based on the type of effects radiations can be classified into ionizing and non – ionizing radiations [109].

Non – Ionizing Radiations

Ultraviolet (UV) rays are the only non-ionizing radiation with mutagenic properties. Their wavelength changes from 100 to 3900 Å. Purines [Adenine (A) or Guanine (G)] and pyrimidines [Cytosine (C) or Thymine (T)] of DNA adsorbing these rays have the maximum absorption at the wavelength of 254 nm [108]. Short wavelength UV rays, can be produced free radicals (-OH), during interaction with water molecules in the cell. Free radicals due to lack of one or more electrons can damage cell proteins or DNA and render it nonfunctional. Except the production of free radicals, UV can directly affect DNA. UV irradiation causes the two T bases fused together through covalenbonds. Such structures are called thymine dimers and cause a distortion in the shape of DNA [109]. This effect is shown in Figure 6 [110].

Ionizing Radiations

Ionizing radiations cause cell damage and can be divided into two groups, namely; particulate and non-particulate radiations. Particulate radiations consist of high energy atom in particles generated due to radioactive decay.

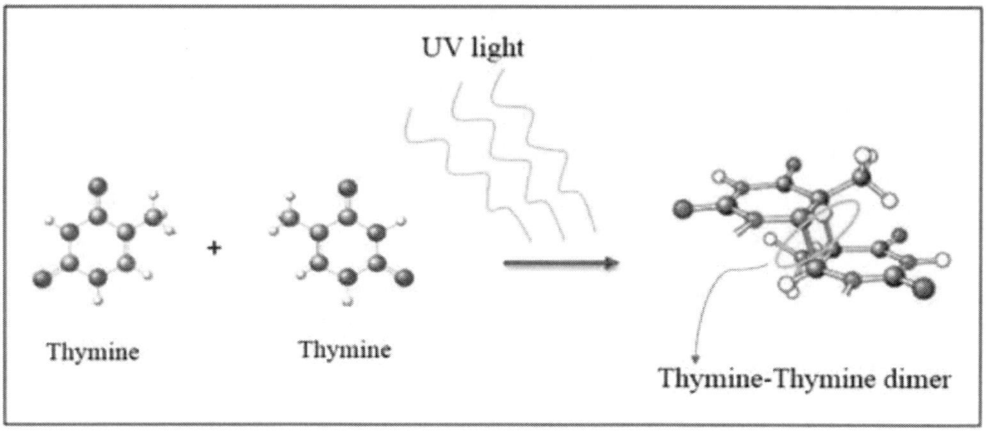

Figure 6. Effect of UV radiation and creation of thymine-thymine bond.

The non-particulate ionizing radiations are represented by X rays and gamma rays which are high energy radiations composed of photons. As same as UV radiation, it seems that the genetic effects of ionizing radiations can be performed directly on DNA or indirect effect via formation of the free radicals [111].

Chemical Mutagens

The first mutagenic action of a chemical was reported by Charlotte Auerbach (1942), who indicated the mutations in cells could be related to the nitrogen mustard. After that, many other mutagenic chemicals have been identified and there is a huge industry and government bureaucracy dedicated to finding them in food additives, industrial wastes, etc. It is possible to distinguish chemical mutagens by their modes of action; some of these cause mutations by mechanisms similar to those which arise spontaneously while others are more like radiation in their effects.

Bromouracil, aminopurine, proflavin, ethidium bromide, nitrous acid, nitrosoguanidine, methyl methanesulfonate and ethyl methanesulphonate are some of the chemical agents which damage or alter the DNA structure [105, 107, 110, 112-114].

ASPERGILLUS AND MUTATION

For efficient bioproducts (e.g., enzymes and organic acids) production, strains can be improved by mutagenesis which is successful method.

Citric acid, fermentation is one of the largest biotechnological industries. Citric acid is a primary product of *Aspergillus niger's* metabolism. According to the latest estimates, citric acid produced about 700,000 tons/annum through fermentation and the need for its productionis ever increasing [115]. The techniques of UV and gamma (γ) ray-induced or chemical mutagenesis have been accepted as routine methods to improve the yield of citric acid by *A. niger* [116].

Table 4. Mutant *Aspergillus spp* products

Strain	Product	Mutagen	Fermentation technique	Results	Ref(s)
A. niger	CMCase, FPase, Xylanase Avicelase	UV	Smf	CMCase activity increased up to two times as compared to the parental strain. FPase activity up to three times. For avicelase and xylanase, UV radiation has slight effect as compared to parental strain.	[114]
A. niger	Phytase	UV	Smf	Mutant strain, phytase (PhytA) activity was 3.3-fold higher than the wild-type activity.	[117]
A. niger	Xylanase, Cellulase	UV	SSF	Cellulase production increased 3.70 times for strain GS1-S059, and 2.65 times for the mutant GS1-S067, as compared to the parental strain. Xylanase production, showed increases of 6.19 and 4.82 times over the parental strain for mutant strains of GS1-S059 and GS1-S067, respectively.	[122]
A. oryzae	B-fructofuranosidase	UV	Smf	Three mutants named IPT-745, IPT-746 and IPT-748 showed the highest amount of mycelium activity whose values increased 1.5 - 1.8 fold, compared with the parent strain	[123]
A. sojae	Polygalacturonase (PG)	UV	SSF	The maximum PG activity in this study was nearly 2 times higher than the activity obtained by Botella et al. [124]	[125]

Strain	Product	Mutagen	Fermentation technique	Results	Ref(s)
A.fumigatus *A.versicolor* *A.nidulans* *A. niger*	Amylase Cellulase	EtBr[1]	Smf	In the cellulase production, almost all the mutated *A.nidulans, A.fumigatus, A.versicolor* and *A.niger* are produced the maximum yielding. In the Amylase production, *A. niger* and *A.nidulans* shows the highest yield in mutated strains.	[126]
A. niger	Cellulase	Hydroxylamine	Smf	Maximal cellulase yield, showed 18% increase in activity respect to wild strain.	[127]
A. niger	Cellulase	Co^{60} γ-rays, UV, Nmethyl-N'-nitro-N nitrosoguanidine	SSF	Mutant strain, produced cellulase in a yield 2.2 fold exceeding that of the wild type.	[128]
A. niger	lipase	UV, nitrous acid	Smf	Mutant strains NAII exhibited 2.53 times more increased lipase activity than the parental strain did.	[129]
A. niger	CMCase FPase β-glucosidase xylanase	UV, γ-rays	SSF	All enzyme activities except CMCase showed 2.5-, 20.4-, 13.1-, and 22.9-fold increases in FPase, β-glucosidase, Xylanase and b-xylosidase, respectively, compared to those obtained in the cultivation of *T. reesei* MCG77[130]	[38]
A. niger	Acid protease	UV, EMS	Smf, SSF	In SSF, 1.53 fold higher acid protease activities with chemical mutant EMS11 over the wild fungal strain. In Smf, UV9 also enhanced the protease production by 2.01 fold over the wild strain.	[119]
A. terreus	Cyclosporin 'A'	UV, EMS	Smf	The highest increment in all mutants was for MC3.2, mutated with EMS, which showed an increase of 885% in drug production.	[131]

[1] Ethidium Bromide.

Xylanase, phytase pectin esterase and pectinase are some of the enzymes which industrially have been produced using mutant *Aspergillus niger* [117-121]. Some products of mutant *Aspergillus spp* are summarized in Table 4.

CONCLUSION

The fungus, *Aspergillus niger* has been considered as one of cellulase enzyme producers. The cost of cellulose production is high due to the slow growth rate of fungi and the low yield of production. The development of economically feasible technologies for cellulase production will facilitate to utilize the large quantities of cellulosic biomass such as the residues of both food industries and agriculture. Producing mutant form of the fungi using UV can improve the yield of cellulose production. Such improved strains can reduce the cost of the process applicable in industrial scales. *Aspergillus niger* has high potential to be mutagenized and genetically modified to obtain an organism capable of producing high levels of cellulases.

REFERENCES

[1] M. J. Taherzadeh and K. Karimi, "Pretreatment of lignocellulosic wastes to improve ethanol and biogas production: a review," *Int. J. Mol. Sci.*, vol. 9, no. 9, pp. 1621–1651, 2008.

[2] S. I. Mussatto and J. A. Teixeira, "Lignocellulose as raw material in fermentation processes," 2010.

[3] P. W. Lambert, J. L. Meers, and D. J. Best, "The Production of Industrial Enzymes [and Discussion]," *Philos. Trans. R. Soc. London. B, Biol. Sci.*, vol. 300, no. 1100, pp. 263–282, 1983.

[4] R. K. Sukumaran, R. R. Singhania, and A. Pandey, "Microbial cellulases-production, applications and challenges," *J. Sci. Ind. Res. (India).*, vol. 64, no. 11, p. 832, 2005.

[5] L. A. Underkofler, R. R. Barton, and S. S. Rennert, "Production of microbial enzymes and their applications," *Appl. Microbiol.*, vol. 6, no. 3, p. 212, 1958.

[6] A. Pandey, P. Selvakumar, C. R. Soccol, and P. Nigam, "Solid state fermentation for the production of industrial enzymes," *Curr. Sci.*, vol. 77, no. 1, pp. 149–162, 1999.

[7] L. M. Prescott, I. P. Harley, and D. A. Klein, "Industrial microbiology," in microbiology, Nyoyorc: MC Graw Hill, 2002, pp. 991–1009.

[8] T. AR and W. PED., Enzymes and their role in biotechnology. London: UK, 1990.

[9] F. Kosikowski and V. V. Mistry, "Cheese and fermented milk foods," 1977.

[10] M. Balandrin, K. JA, W. ES, and B. WH, "Natural plant chemicals: sources of industrial and medicinal materials.," *Science (80-.).*, vol. 228, pp. 1154–60, 1985.

[11] A. Illanes, Enzyme biocatalysis: principles and applications. Springer Science and Business Media, 2008.

[12] D. S. Clark and H. W. Blanch, Biochemical engineering. CRC Press, 1997.

[13] P. S. Nigam, "Microbial enzymes with special characteristics for biotechnological applications," *Biomolecules*, vol. 3, no. 3, pp. 597–611, 2013.

[14] D. Neto Ferreira Gomes and A. de Queiroz Cavalcanti, "Diversidade e potencial biotecnólogico de fungos filamentosos isolados do manguezal Barra das Jangadas, Jaboatão dos Guararapes, Pernambuco," 2007.

[15] V. M. Taragano and A. M. R. Pilosof, "Application of Doehlert designs for water activity, pH, and fermentation time optimization for *Aspergillus niger* pectinolytic activities production in solid-state and submerged fermentation," *Enzyme Microb. Technol.*, vol. 25, no. 3, pp. 411–419, 1999.

[16] M. Uenojo and G. M. Pastore, "Pectinases: aplicações industriais e perspectivas," *Quim. Nova*, vol. 30, no. 2, p. 388, 2007.

[17] B. De Paula, I. V. M. Moraes, C. C. Castilho, F. D. O. S. S. Gomes, V. M. Da Matta, and L. M. C. Cabral, "Melhoria na eficiência da clarificação de suco de maracujá pela combinação dos processos de microfiltração e enzimático," *Bol. do Cent. Pesqui. Process. Aliment.*, vol. 22, no. 2, 2004.

[18] A. De Gregorio, G. Mandalari, N. Arena, F. Nucita, M. M. Tripodo, and R. B. Lo Curto, "SCP and crude pectinase production by slurry-state fermentation of lemon pulps," *Bioresour. Technol.*, vol. 83, no. 2, pp. 89–94, 2002.

[19] A. Pandey, S. Benjamin, C. R. Soccol, P. Nigam, N. Krieger, and V. T. Soccol, "The realm of microbial lipases in biotechnology," *Biotechnol. Appl. Biochem.*, vol. 29, no. 2, pp. 119–131, 1999.

[20] C. R. Soccol and L. P. S. Vandenberghe, "Overview of applied solid-state fermentation in Brazil," *Biochem. Eng. J.*, vol. 13, no. 2, pp. 205–218, 2003.

[21] A. Aquino, J. A. Jorge, H. F. Terenzi, and M. Polizeli, "Studies on a thermostable α-amylase from the thermophilic fungus Scytalidium thermophilum," *Appl. Microbiol. Biotechnol.*, vol. 61, no. 4, pp. 323–328, 2003.

[22] Q. D. Nguyen, J. M. Rezessy-Szabó, M. Claeyssens, I. Stals, and Á. Hoschke, "Purification and characterisation of amylolytic enzymes from thermophilic fungus Thermomyces lanuginosus strain ATCC 34626," *Enzyme Microb. Technol.*, vol. 31, no. 3, pp. 345–352, 2002.

[23] S. C. Peixoto, J. A. Jorge, H. F. Terenzi, and T. M. Maria de Lourdes, "Rhizopus microsporus var. rhizopodiformis: a thermotolerant fungus with potential for production of thermostable amylases," *Int. Microbiol.*, vol. 6, no. 4, pp. 269–273, 2003.

[24] S. Sivaramakrishnan, D. Gangadharan, K. M. Nampoothiri, C. R. Soccol, and A. Pandey, "a-Amylases from microbial sources–an overview on recent developments," *Food Technol. Biotechnol.*, vol. 44, no. 2, pp. 173–184, 2006.

[25] N. D. Rawlings and A. J. Barrett, "Families of serine peptidases.," *Methods Enzymol.*, vol. 244, p. 19, 1994.

[26] C. G. Kumar, "Purification and characterization of a thermostable alkaline protease from alkalophilic Bacillus pumilus," *Lett. Appl. Microbiol.*, vol. 34, no. 1, pp. 13–17, 2002.

[27] R. Rahman, M. Basri, and A. B. Salleh, "Thermostable alkaline protease from Bacillus stearothermophilus F1; nutritional factors affecting protease production," *Ann. Microbiol.*, vol. 53, no. 2, pp. 199–210, 2003.

[28] A. Gupta, B. Joseph, A. Mani, and G. Thomas, "Biosynthesis and properties of an extracellular thermostable serine alkaline protease from Virgibacillus pantothenticus," *World J. Microbiol. Biotechnol.*, vol. 24, no. 2, pp. 237–243, 2008.

[29] T. Robinson, B. Chandran, and P. Nigam, "Studies on the production of enzymes by white-rot fungi for the decolourisation of textile dyes," *Enzyme Microb. Technol.*, vol. 29, no. 8, pp. 575–579, 2001.

[30] P. Dwivedi, V. Vivekanand, N. Pareek, A. Sharma, and R. P. Singh, "Bleach enhancement of mixed wood pulp by xylanase–laccase concoction derived through co-culture strategy," *Appl. Biochem. Biotechnol.*, vol. 160, no. 1, pp. 255–268, 2010.

[31] G. N. Kumar and K. Srikumar, "Characterization of xerophytic thermophilic laccase exhibiting metal ion-dependent dye decolorization potential," *Appl. Biochem. Biotechnol.*, vol. 167, no. 3, pp. 662–676, 2012.

[32] L. Papinutti, P. Dimitriu, and F. Forchiassin, "Stabilization studies of Fomes sclerodermeus laccases," *Bioresour. Technol.*, vol. 99, no. 2, pp. 419–424, 2008.

[33] S. Wongwilaiwalin, U. Rattanachomsri, T. Laothanachareon, L. Eurwilaichitr, Y. Igarashi, and V. Champreda, "Analysis of a thermophilic lignocellulose degrading microbial consortium and multi-species lignocellulolytic enzyme system," *Enzyme Microb. Technol.*, vol. 47, no. 6, pp. 283–290, 2010.

[34] M. K. Bhat, "Cellulases and related enzymes in biotechnology," *Biotechnol. Adv.*, vol. 18, no. 5, pp. 355–383, 2000.

[35] M. K. Bhat and S. Bhat, "Cellulose degrading enzymes and their potential industrial applications," *Biotechnol. Adv.*, vol. 15, no. 3, pp. 583–620, 1997.

[36] M. S. Chandra, B. Viswanath, and B. R. Reddy, "Cellulolytic enzymes on lignocellulosic substrates in solid state fermentation by *Aspergillus niger*," *Indian J. Microbiol.*, vol. 47, no. 4, pp. 323–328, 2007.

[37] T. V. Ojumu, B. O. Solomon, E. Betiku, S. K. Layokun, and B. Amigun, "Cellulase Production by *Aspergillus flavus* Linn Isolate NSPR 101 fermented in sawdust, bagasse and corncob," *African J. Biotechnol.*, vol. 2, no. 6, pp. 150–152, 2003.

[38] S. W. Kang, Y. S. Park, J. S. Lee, S. I. Hong, and S. W. Kim, "Production of cellulases and hemicellulases by *Aspergillus niger* KK2 from lignocellulosic biomass," *Bioresour. Technol.*, vol. 91, no. 2, pp. 153–156, 2004.

[39] P. B. Acharya, D. K. Acharya, and H. A. Modi, "Optimization for cellulase production by *Aspergillus niger* using saw dust as substrate," *African J. Biotechnol.*, vol. 7, no. 22, 2008.

[40] M. Sohail, R. Siddiqi, A. Ahmad, and S. A. Khan, "Cellulase production from *Aspergillus niger* MS82: effect of temperature and pH," *N. Biotechnol.*, vol. 25, no. 6, pp. 437–441, 2009.

[41] P. Nigam and K. A. Prabhu, "Effect of cultural factors on cellulase biosynthesis in submerged bagasse fermentation by basidiomycetes cultures," *J. Basic Microbiol.*, vol. 31, no. 4, pp. 285–292, 1991.

[42] Y. Park, S. Kang, J. Lee, S. L. Hong, and S. Kim, "Xylanase production in solid state fermentation by *Aspergillus niger* mutant using statistical experimental designs," *Appl. Microbiol. Biotechnol.*, vol. 58, no. 6, pp. 761–766, 2002.

[43] A. Archana and T. Satyanarayana, "Xylanase production by thermophilic Bacillus licheniformis A99 in solid-state fermentation," *Enzyme Microb. Technol.*, vol. 21, no. 1, pp. 12–17, 1997.

[44] Q. K. Beg, B. Bhushan, M. Kapoor, and G. S. Hoondal, "Enhanced production of a thermostable xylanase from Streptomyces sp. QG-11-3 and its application in

biobleaching of eucalyptus kraft pulp," *Enzyme Microb. Technol.*, vol. 27, no. 7, pp. 459–466, 2000.

[45] S. Q. Yang, Q. J. Yan, Z. Q. Jiang, L. T. Li, H. M. Tian, and Y. Z. Wang, "High-level of xylanase production by the thermophilic Paecilomyces themophila J18 on wheat straw in solid-state fermentation," *Bioresour. Technol.*, vol. 97, no. 15, pp. 1794–1800, 2006.

[46] M. Polizeli, A. C. S. Rizzatti, R. Monti, H. F. Terenzi, J. A. Jorge, and D. S. Amorim, "Xylanases from fungi: properties and industrial applications," *Appl. Microbiol. Biotechnol.*, vol. 67, no. 5, pp. 577–591, 2005.

[47] N. R. Kamini, J. G. S. Mala, and R. Puvanakrishnan, "Lipase production from *Aspergillus niger* by solid-state fermentation using gingelly oil cake," *Process Biochem.*, vol. 33, no. 5, pp. 505–511, 1998.

[48] A. K. Gombert, A. L. Pinto, L. R. Castilho, and D. M. G. Freire, "Lipase production by Penicillium restrictum in solid-state fermentation using babassu oil cake as substrate," *Process Biochem.*, vol. 35, no. 1, pp. 85–90, 1999.

[49] N. D. Mahadik, U. S. Puntambekar, K. B. Bastawde, J. M. Khire, and D. V Gokhale, "Production of acidic lipase by *Aspergillus niger* in solid state fermentation," *Process Biochem.*, vol. 38, no. 5, pp. 715–721, 2002.

[50] J. Shendure and H. Ji, "Next-generation DNA sequencing.," *Nat. Biotechnol.*, vol. 26, no. 10, pp. 1135–1145, 2008.

[51] H. Treichel, D. de Oliveira, M. A. Mazutti, M. Di Luccio, and J. V. Oliveira, "A review on microbial lipases production," *Food Bioprocess Technol.*, vol. 3, no. 2, pp. 182–196, 2010.

[52] R. Kaushal, N. Sharma, and D. Tandon, "Cellulase and xylanase production by co-culture of *Aspergillus niger* and Fusarium oxysporum utilizing forest waste," *Turk. J. Biochem.*, vol. 37, pp. 35–41, 2012.

[53] T. H. Pham, D. T. Quyen, and N. M. Nghiem, "Purification and properties of an endoglucanase from *Aspergillus niger VTCC-F021*," *Turkish J. Biol.*, vol. 36, no. 6, pp. 694–701, 2012.

[54] X. Li, H. Yang, B. Roy, D. Wang, W. Yue, L. Jiang, E. Y. Park, and Y. Miao, "The most stirring technology in future: Cellulase enzyme and biomass utilization," *African J. Biotechnol.*, vol. 8, no. 11, 2009.

[55] S. Mrudula and R. Murugammal, "Production of cellulase by *Aspergillus niger* under submerged and solid state fermentation using coir waste as a substrate," *Brazilian J. Microbiol.*, vol. 42, no. 3, pp. 1119–1127, 2011.

[56] I. Soares, R. P. Barcelos, S. Baroni, and Z. Távora, Microorganism-Produced Enzymes in the Food Industry. INTECH Open Access Publisher, 2012.

[57] R. C. Kuhad, R. Gupta, and A. Singh, "Cellulase Production by Utilizing Agricultural Wastes," *Enzyme Res.*, vol. 2011, 2011.

[58] J. A. Khan and S. K. Singh, "Production of cellulase using cheap substrates by solid state fermentation," *Int. J. Plant Anim. Environ. Sci.*, vol. 1, pp. 179–187, 2011.

[59] K. M. Wong, N. Abdul Rahman, S. Vikineswary, and M. A. Hassan, "Enzymatic hydrolysis of palm oil mill effluent solid using mixed cellulases from locally isolated fungi.," *Res. J. Microbiol.*, vol. 3, no. 6, pp. 474–481, 2008.

[60] P. F. Omojasola and O. P. Jilani, "Cellulase production by Trichoderma longi, *Aspergillus niger and Saccharomyces cerevisae* cultured on waste materials from orange," *Pakistan J. Biol. Sci.*, vol. 11, no. 20, pp. 2382–2388, 2008.

[61] P. F. Omojasola and O. P. Jilani, "Cellulase production by Trichoderma longi, *Aspergillus niger and Saccharomyces cerevisae* cultured on plantain peel," *Res. J. Microbiol.*, vol. 4, no. 2, pp. 67–74, 2009.

[62] T. Juhász, K. Kozma, and K. Réczey, "Production of β-glucosidase in mixed culture of *Aspergillus niger BKMF 1305 and Trichoderma reesei RUT C30*," *Food Technol. Biotechnol.*, vol. 41, no. 1, pp. 49–53, 2003.

[63] C. Pasha, B. C. Sekhar, B. Srinivas, K. Balakrishna, and N. Hanumalal, "Sequential cellulase production, saccharification and ethanol fermentation using rice straw," *J. Sci. Ind. Res. (India).*, vol. 71, pp. 616–620, 2012.

[64] M. C. Devi and M. S. Kumar, "Production, Optimization and Partial purification of Cellulase by *Aspergillus niger* fermented with paper and timber sawmill industrial wastes," *J. Microbiol. Biotechnol. Res.*, vol. 2, no. 1, pp. 120–128, 2012.

[65] M. U. Kiranmayi, S. Poda, P. Charyulu, M. Vijayalakshmi, and P. V Krishna, "Studies on influence of natural biowastes on cellulase production by *Aspergillus niger*," *J. Environ. Biol.*, vol. 32, no. 6, 2011.

[66] H. A. Murad and H. E. H. Azzaz, "Cellulase Production from Rice Straw by *Aspergillus flavus NRRL 552*," *Sci. Int.*, vol. 1, no. 4, pp. 103–107, 2013.

[67] I. F. Okonkwo, F. J. C. Odibo, and C. M. Obele, "Time course for cellulase enzyme production by *Aspergillus Flavus* using different organic nitrogen sources," *Sci. J. Biol. Sci.*, vol. 3, no. 2, pp. 24–28, 2014.

[68] G. A. Youssef and M. M. Berekaa, "Improved production of endoglucanase enzyme by *Aspergillus terreus*; application of Plackett-Burman design for optimization of process parameters.," *Biotechnology*, vol. 8, no. 2, pp. 212–219, 2009.

[69] M. Shahriarinour, M. N. A. Wahab, R. Mohamad, S. Mustafa, and A. B. Ariff, "Effect of medium composition and cultural condition on cellulase production by *Aspergillus terreus*," *African J. Biotechnol.*, vol. 10, no. 38, pp. 7459–7467, 2013.

[70] J. A. R., G. A. V., M. S. M., M. S. B., and S. Khan, "Cellulase Production by Utilizing Agricultural Wastes," *Res. J. Agric. For. Sci.*, vol. 1(7), pp. 6–9, 2013.

[71] M. Prabakaran, V. Thennarasu, R. A. Mangala, R. Bharathidasan, N. Chandrakala, and N. Mohan, "Comparative studies on the enzyme activities of wild and mutant fungal strains isolated from sugarcane field," *Indian J. Sci. Technol.*, vol. 2, no. 11, pp. 46–49, 2009.

[72] A. A. Sherief, A. B. El-Tanash, and N. Atia, "Cellulase production by *Aspergillus fumigatus* grown on mixed substrate of rice straw and wheat bran," 2010.

[73] U. H. Ikram-ul-Haq, K. Shahzadi, M. M. Javed, S. Ali, and M. A. Qadeer, "Cotton saccharifying activity of cellulases by Trichoderma harzianum UM-11 in shake flask," *Int. J. Bot.*, vol. 1, no. 1, pp. 19–22, 2005.

[74] J. Medve, J. Ståhlberg, and F. Tjerneld, "Isotherms for adsorption of cellobiohydrolase I and II fromtrichoderma reesei on microcrystalline cellulose," *Appl. Biochem. Biotechnol.*, vol. 66, no. 1, pp. 39–56, 1997.

[75] J. Weber and F. A. Agblevor, "Microbubble fermentation of Trichoderma reesei for cellulase production," *Process Biochem.*, vol. 40, no. 2, pp. 669–676, 2005.

[76] G. Zou, S. Shi, Y. Jiang, J. van den Brink, R. P. de Vries, L. Chen, J. Zhang, L. Ma, C. Wang, and Z. Zhou, "Construction of a cellulase hyper-expression system in Trichoderma reesei by promoter and enzyme engineering," *Microb. Cell Fact.*, vol. 11, no. 1, p. 21, 2012.

[77] G. Narasimha, A. Sridevi, V. Buddolla, C. M. Subhosh, and R. B. Rajasekhar, "Nutrient effects on production of cellulolytic enzymes by *Aspergillus niger*," 2006.

[78] M. Gruno, P. Väljamäe, G. Pettersson, and G. Johansson, "Inhibition of the Trichoderma reesei cellulases by cellobiose is strongly dependent on the nature of the substrate," *Biotechnol. Bioeng.*, vol. 86, no. 5, pp. 503–511, 2004.

[79] N. Aro, M. Ilmén, A. Saloheimo, and M. Penttilä, "ACEI of Trichoderma reesei is a repressor of cellulase and xylanase expression," *Appl. Environ. Microbiol.*, vol. 69, no. 1, pp. 56–65, 2003.

[80] M. Mandels and E. T. Reese, "Induction of cellulase in Trichoderma viride as influenced by carbon sources and metals," *J. Bacteriol.*, vol. 73, no. 2, p. 269, 1957.

[81] G. Beldman, M. F. Leeuwen, F. M. Rombouts, and F. G. J. Voragen, "The cellulase of Trichoderma viride," *Eur. J. Biochem.*, vol. 146, no. 2, pp. 301–308, 1985.

[82] L. E. Berghem and L. G. Pettersson, "The mechanism of enzymatic cellulose degradation. Purification of a cellulolytic enzyme from Trichoderma viride active on highly ordered cellulose.," *Eur. J. Biochem.*, vol. 37, no. 1, p. 21, 1973.

[83] J. Liu, X. Yuan, G. Zeng, J. Shi, and S. Chen, "Effect of biosurfactant on cellulase and xylanase production by Trichoderma viride in solid substrate fermentation," *Process Biochem.*, vol. 41, no. 11, pp. 2347–2351, 2006.

[84] A. S. Lakshmi and G. Narasimha, "Production of cellulases by fungal cultures isolated from forest litter soil," *Ann. For. Res.*, vol. 55, no. 1, pp. 85–92, 2012.

[85] M. G. Adsul, K. B. Bastawde, A. J. Varma, and D. V. Gokhale, "Strain improvement of Penicillium janthinellum NCIM 1171 for increased cellulase production," *Bioresour. Technol.*, vol. 98, no. 7, pp. 1467–1473, 2007.

[86] M. Sohail, A. Ahmad, and S. A. Khan, "Production of cellulases from Alternaria sp. MS28 and their partial characterization," *Pak. J. Bot.*, vol. 43, no. 6, pp. 3001–3006, 2011.

[87] A. J. M. Reddi Pradeep, A. P. Kumar, and G. Narasimha, "Induction of chemical mutations in *Aspergillus niger* to enhance cellulase production," *Int. J. Environ. Biol.*, vol. 2(3), pp. 129–132, 2012.

[88] S. Sadhu and T. K. Maiti, "Cellulase production by bacteria: a review," *Br. Microbiol. Res. J.*, vol. 3, no. 3, pp. 235–258, 2013.

[89] I. Persson, F. Tjerneld, and B. Hahn-Hägerdal, "Fungal cellulolytic enzyme production: a review," *Process Biochem.*, vol. 26, no. 2, pp. 65–74, 1991.

[90] K. R. Sharrock, "Cellulase assay methods: a review," *J. Biochem. Biophys. Methods*, vol. 17, no. 2, pp. 81–105, 1988.

[91] B. Wu, Y. Zhao, and P. J. Gao, "A new approach to measurement of saccharifying capacities of crude cellulase," *Bio. Resources*, vol. 1, no. 2, pp. 189–200, 2007.

[92] Y.-H. P. Zhang, M. E. Himmel, and J. R. Mielenz, "Outlook for cellulase improvement: screening and selection strategies," *Biotechnol. Adv.*, vol. 24, no. 5, pp. 452–481, 2006.

[93] M. Dashtban, M. Maki, K. T. Leung, C. Mao, and W. Qin, "Cellulase activities in biomass conversion: measurement methods and comparison," *Crit. Rev. Biotechnol.*, vol. 30, no. 4, pp. 302–309, 2010.

[94] M. Mandels, R. Andreotti, and C. Roche, "Measurement of saccharifying cellulase," in Biotechnol. Bioeng. Symp.;(United States), 1976, vol. 6.

[95] C. P. Kubicek, "beta-Glucosidase excretion by Trichoderma pseudokoningii: correlation with cell wall bound beta-1.3-glucanase activities.," *Arch. Microbiol.*, vol. 132(4), no. 1982, pp. 349–354.

[96] M. Irfan, U. Irfan, Z. Razaq, Q. Syed, and M. Nadeem, "Utilization of agricultural wastes as a substrate for Carboxymethyl cellulase production from *Aspergillus niger* in submerged fermentation," *Int. J. Agro Vet. Med. Sci.*, vol. 5, no. 5, pp. 464–471, 2011.

[97] S. Baig and M. Saleem, "Production and characterization of cellulases of *Aspergillus niger* by using rice husk and saw dust as substrates," *Pakistan J. Bot.*, 2012.

[98] F. Abubakar and O. Oloyede, "Production and activity of cellulase from *Aspergillus niger* using rice bran and orange peel as substrates," *Int. J. Sci. Res. Manag.*, vol. 1, no. (5), pp. 285–291, 2013.

[99] S. N. Chinedu and V. I. Okochi, "Cellulase Production by Wild-type *Aspergillus niger*, Penicillium chrysogenum and Trichoderma harzianum Using Waste Cellulosic Materials," Unspecified, 2010.

[100] A. Vyas, D. Vyas, and K. M. Vyas, "Production and optimization of cellulase on pretreated groundnut shell by *Aspergillus terreusAV49.*," *Sci. Ind. Res.*, 2005; 64281-286., vol. 64, pp. 281–286, 2005.

[101] M. Kumakura, "Preparation of immobilized cellulase beads and their application to hydrolysis of cellulosic materials," *Process Biochem.*, vol. 32, no. 7, pp. 555–559, 1997.

[102] L. Fan, M. M. Gharpuray, Y.-H. Lee, S. Aiba, A. Fiechter, J. Klein, and K. Schügerl, *Cellulose hydrolysis*. Springer-verlag Berlin, 1987.

[103] P. Chand, A. Aruna, A. M. Maqsood, and L. V Rao, "Novel mutation method for increased cellulase production," *J. Appl. Microbiol.*, vol. 98, no. 2, pp. 318–323, 2005.

[104] J. Barrios-Gonzalez, F. J. Fernandez, and A. Tomasini, "Microbial secondary metabolites production and strain improvement," *Indian J. Biotechnol.*, vol. 2, no. 3, pp. 322–333, 2003.

[105] H. F. Lodish, A. Berk, S. L. Zipursky, P. Matsudaira, D. Baltimore, and J. Darnell, "essentials of genetic," in Text book of medical biochemistry, Second., vol. 4, Elsevier, 2006, p. 578.

[106] J. L. M. Coimbra, F. I. F. de Carvalho, and A. C. de Oliveira, "Genetic variability induced by chemical and physical mutagenic agents in oat genotypes," *Crop Breed. Appl. Biotechnol.*, vol. 4, no. 1, pp. 1–7, 2004.

[107] P. Khanna, "Essentials of Genetics," in *Essentials of Genetics*, New delhi: IK International Pvt Ltd, 2010, pp. 255–269.

[108] H. Ikehata and T. Ono, "The mechanisms of UV mutagenesis," *J. Radiat. Res.*, vol. 52, no. 2, pp. 115–125, 2011.

[109] R. E. Johnson, S. Prakash, and L. Prakash, "Efficient bypass of a thymine-thymine dimer by yeast DNA polymerase, Polη," *Science (80).*, vol. 283, no. 5404, pp. 1001–1004, 1999.

[110] M. B. Habibi Najafi, "Bacterial Mutation;; Types, Mechanisms and Mutant Detection Methods: A Review," in 1st Global Multidisciplinary eConference, 2013.

[111] B. S. Ahloowalia and M. Maluszynski, "Induced mutations–A new paradigm in plant breeding," *Euphytica*, vol. 118, no. 2, pp. 167–173, 2001.

[112] M. Jobling, M. Hurles, and C. Tyler-Smith, "The diversity of human genome," in Human evolutionary genetics: origins, peoples and disease, Newyork: Garland Science, 2013, p. 52,53.

[113] A. Kodym and R. Afza, "Physical and chemical mutagenesis," in Plant Functional Genomics, Springer, 2003, pp. 189–203.

[114] M. Irfan, J. Javed, and Q. Syed, "UV mutagenesis of *Aspergillus niger* for enzyme production in submerged fermentation," *Pakistan J. Biochem. Mol. Biol.*, vol. 44, no. 4, pp. 137–140, 2011.

[115] A. Shoukat, M. A. Ahmed, and G. R. Akbar, "Citric acid production by *Aspergillus niger* by batch cultivation," *J. Biotechnol.*, vol. 7, pp. 47–53, 1997.

[116] S. Khurshid, S. Ali, H. Ashraf, M. A. Qadeer, and M. I. Rajoka, "Mutation of *Aspergillus niger* for hyperproduction of citric acid from black strap molasses," *World J. Microbiol. Biotechnol.*, vol. 17, no. 1, pp. 35–37, 2001.

[117] M. K. Chelius and R. J. Wodzinski, "Strain improvement of *Aspergillus niger* for phytase production," *Appl. Microbiol. Biotechnol.*, vol. 41, no. 1, pp. 79–83, 1994.

[118] P. Antier, A. Minjares, S. Roussos, M. Raimbault, and G. Viniegra-Gonzalez, "Pectinase-hyperproducing mutants of *Aspergillus niger* C28B25 for solid-state fermentation of coffee pulp," *Enzyme Microb. Technol.*, vol. 15, no. 3, pp. 254–260, 1993.

[119] S. Radha, R. H. Babu, A. Sridevi, N. B. L. Prasad, and G. Narasimha, "Development of mutant fungal strains of *Aspergillus niger* for enhanced production of acid protease in submerged and solid state fermentation," *Eur. J. Exp. Biol.*, vol. 2, no. 5, pp. 1517–1528, 2012.

[120] S. W. Kang, E. H. Ko, J. S. Lee, and S. W. Kim, "Over-production of β-glucosidase by *Aspergillus niger* mutant from lignocellulosic biomass," *Biotechnol. Lett.*, vol. 21, no. 8, pp. 647–650, 1999.

[121] M. R. Roja and N. N. Prasad, "Strain improvement of *aspergillus flavus* for enhanced production of alkaline protease enzyme," *Int. J. Sc. Innov. Discov.*, vol. 2, no. 1, pp. 113–120, 2012.

[122] D. Nicolás-Santiago, C. Regalado-González, B. García-Almendárez, F. J. Fernández, A. Téllez-Jurado, and S. Huerta-Ochoa, "Physiological, morphological, and mannanase production studies on *Aspergillus niger* uam-gs1 mutants," *Electron. J. Biotechnol.*, vol. 9, no. 1, p. 0, 2006.

[123] B. G. Maresma, B. G. Castillo, R. C. Fernández, E. S. da Silva, A. E. Maiorano, and M. F. de A. Rodrigues, "Mutagenesis of *Aspergillus oryzae IPT-301* to improve the production of β-fructofuranosidase," *Brazilian J. Microbiol.*, vol. 41, no. 1, pp. 186–195, 2010.

[124] C. Botella, I. De Ory, C. Webb, D. Cantero, and A. Blandino, "Hydrolytic enzyme production by *Aspergillus awamori* on grape pomace," *Biochem. Eng. J.*, vol. 26, no. 2, pp. 100–106, 2005.

[125] H. Demir, N. Gogus, C. Tari, D. Heerd, and M. F. Lahore, "Optimization process parameters for the utilization of orange peel to produce polygalacturonase by solid-state fermentation from an *Aspergillus sojae* mutant strain," *Turk. J. Biol.*, vol. 36, pp. 394–404, 2012.

[126] G. Sathyaprabha, A. Panneerselvam, and S. Muthukkumarasamy, "Production of Cellulase and Amylase from wild and mutated fungal isolates," *E-Journal Life Sci.*, vol. 1, no. 1, pp. 39–45, 2011.

[127] D. Damisa, F. A. Kuta, and N. U. Adabara, "Mutagenic treatment of *Aspergillus niger* with hydroxylamine for improved cellulase synthesis from cellulosic wastes," *J. Biotechnol. Sci.*, vol. 1, pp. 65–72, 2013.

[128] V. H. Vu, T. A. Pham, and K. Kim, "Improvement of fungal cellulase production by mutation and optimization of solid state fermentation," *Mycobiology*, vol. 39, no. 1, pp. 20–25, 2011.

[129] J. G. Sandana Mala, N. R. Kamini, and R. Puvanakrishnan, "Strain improvement of *Aspergillus niger* for enhanced lipase production.," *J. Gen. Appl. Microbiol.*, vol. 47, no. 4, pp. 181–186, 2001.

[130] P. J. Considine, A. O'Rorke, T. J. Hackett, and M. P. Coughlan, "Hydrolysis of beet pulp polysaccharides by extracts of solid-state cultures of *Pencicillium capsulatum*," *Biotechnol. Bioeng.*, vol. 31, no. 5, pp. 433–438, 1988.

[131] W. Irum and T. Anjum, "Production enhancement of Cyclosporin 'A'by *Aspergillus terreus* through mutation," *African J. Biotechnol.*, vol. 11, no. 7, pp. 1736–1743, 2014.

About the Editors

Dr. Aydin Berenjian
Assistant Professor
The University of Waikato, New Zealand
aydinb@waikato.ac.nz

Dr. Hoda Jafarizadeh-Malmiri
Assistant Professor
Sahand University of Technology, Iran
h_jafarizadeh@sut.ac.ir

Dr. Yuanda Song
Professor
Shandong University of Technology, China
ysong@sdut.edu.cn

INDEX

C

F

G

H

I

J

K

L

M

N

Q

R

S

T

U

V

W

X

Y

Z